MW00527820

Green Humour for a Greying Planet

ADVANCE PRAISE FOR THE BOOK

'Rohan speaks the truth to power and the truth to our conscience. He is gifted with the ability to convey hard truths about science and people with a skill that is rare and unique. I look out for his work because it always compels my growth and understanding. This compilation of his works is as important as the urgent work necessary to act on climate change and restore ecosystems. A must-read!'—Dia Mirza, actor, producer, United Nations SDGs advocate, Wildlife Trust of India ambassador

'I have personally been an avid follower of Green Humour on social media for several years. Rohan finds ways to portray the most stark environmental injustices in a single comic strip with a brand of humour that is both funny and tragic. I am yet to come across a more talented and effective advocate for Mother Earth and all her voiceless children'—Faye D'Souza, journalist and entrepreneur

Green Humour for a Greying Planet

ROHAN CHAKRAVARTY

PENGUIN BOOKS
An imprint of Penguin Random House

PENGUIN BOOKS

USA | Canada | UK | Ireland | Australia
New Zealand | India | South Africa | China

Penguin Books is part of the Penguin Random House group of companies whose addresses can be found at global.penguinrandomhouse.com

Published by Penguin Random House India Pvt. Ltd
4th Floor, Capital Tower 1, MG Road,
Gurugram 122 002, Haryana, India

Penguin Random House India

First published in Penguin Books by Penguin Random House India 2021

10 9 8 7 6 5 4 3

ISBN 9780143452959

Typeset in Adobe Garamond Pro by Manipal Technologies Limited, Manipal
Printed at Replika Press Pvt. Ltd, India

www.penguin.co.in

For the two tigresses in my life:
my late mother, Sulabha;
and the first wild tigress I ever saw,
who lured me into drawing cartoons on wildlife and conservation

MAHUA ESSENCE SHOWER GEL

Contents

Foreword

Over the years, we have written for magazines, newspapers and scientific journals, most of which was pretty dry and serious. But now we are more convinced than ever that the way to approach subjects like wildlife conservation, human–animal conflict and the downward spiral we humans seem to be bent on taking our green planet on is through humour, more specifically, green humour.

For far too long, conversations around these subjects have been serious business. All gloom and doom. But there's a place for humour, too. A sense of timing, the set-up and the punch line are important. Until now, not many people were proficient or confident about pulling it off. It's not just the subject that needs levity. Even the people practising conservation take themselves far too seriously. We've long needed witty cartoons that target both the subject as well as the practitioners.

Well, India has it now, in the shape of Rohan's new book. Yep, it has the phrase 'green humour' in its title. Finally, we have a way to help fight the uphill struggle and a means to keep people smiling while they do it!

Rohan uses wit, his considerable knowledge of wildlife-related issues and a huge array of comic characters from his artistic repertoire to make us chuckle. Recently, one of us had to give a talk on frogs to distinguished colleagues in New Delhi. Well, being a reptile person it was extremely helpful to be able to use a couple of Rohan's frog cartoons to spice up my PowerPoint presentation, which could otherwise have been deadening. In one cartoon, a toad was trying to convince a snake not to eat him because he tasted nasty and was warty. The snake responded with this: 'No problem, I've got ketchup.'

Whether it's the subject of climate change, human–animal conflict, hunting, poaching or wildlife science, Rohan always finds a comic character with something bizarre, ironic and/or funny to say about it. In this day of self-important bureaucrats, scientists and know-it-alls, it's most refreshing to have laughter come to the rescue!

—Romulus Whitaker and Janaki Lenin

Introduction

Ask avid newspaper readers about which politician took a jibe at who in the Parliament and they'll be ready with an answer and the entire quote. Ask subscribers of news apps who wore which outlandish dress at Cannes and they'll be able to tell you the price and the name of the designer. Now ask them which species of snake was discovered recently and you'll be greeted with a puzzled silence. Why? Because wildlife, unlike politicians and celebrities, rarely makes it to Page 1 or Page 3. And because wildlife generally cannot stage scandals and wardrobe malfunctions to make the headlines, the task of bridging that knowledge gap falls on creative visual communication.

'Cartooning is a barometer of freedom,' says exiled Venezuelan cartoonist Rayma Suprani. Throughout history, cartoons have been every reader's go-to section in the newspaper, every subscriber's favourite window on his/her app. While in newspapers the cartoon has always done the job of reflecting the mood of the paper, the reader and the nation on any given day; online, the art form has become one of the most 'shareable' commodities, serving as a humorous summary of information and daily events. But environmental matters, by and large, have been missing from this canvas.

American cartoonist Jay Norwood 'Ding' Darling had got that ball rolling in the 1960s, with cartoons on unsustainable game hunting and species extinction frequenting his usually political column. Finnish cartoonist and naturalist Seppo Leinonen carried that baton forward, with his entire cartooning portfolio devoted to natural history and environmental issues, made all the merrier with his proficient craftsmanship. Canadian biologist-cum-cartoonist Rosemary Mosco brings in a very fresh and contemporary twist to the genre with her quirky humour and adorable characters. In India, the wildlife cartooning bug bit yours truly when I crossed paths with my first wild tigress in the Nagzira reserve, a few hours away from Nagpur, my home town.

Having dabbled as a cartoonist, trying out everything from aliens to social commentary, I found my feet only when I considered wildlife as my muse.

And thankfully the relationship has proven to be symbiotic. In the ten years that my series, Green Humour, has existed, been published and read across different media, my cartoons have served readers of all ages, interests, ideologies and vocations as informational snippets on environmental happenings. What started as something for my own satisfaction and amusement gradually assumed the role of the friendly neighbourhood mediator between conservation science, biology and the layman. Green Humour started off with sporadic appearances in the ace wildlife magazine *Sanctuary Asia*, later acquiring columns with *Saevus* magazine, *Tinkle Digest* and *Sustainuance* magazine, among others. The Universal Press Syndicate picked up the series for online syndication in 2013 on its webcomics channel Gocomics. Since then, Green Humour has run periodically in BL Ink, *Sunday Mid-Day*, *Pune Mirror* and *The Hindu*. Through the course of developing this series, I have grown to realize that cartoons on conservation work in three ways: deliver the message of conservation without making it preachy, eliminate jargon and make the information being presented easy to retain and respond to, and instil a curiosity and respect for the natural world in the mind of the reader. It is this friendly handshake between the natural world and our minds that I invite you to make by reading this book.

ONE

Climate Change and **Ecological Imbalance**

THANKS TO CLIMATE CHANGE, WE'VE PERMANENTLY DISPLACED THE MOST POPULAR CARTOON CLICHÉ.
#1

Thanks to capitalism and industrialization, the last decade witnessed continued destruction of nature and its resources. World leaders, despite being signatories to the Kyoto Protocol, Paris Agreement and You-Name-It-treaties, have consistently failed to keep their green promises. And while the media has always found environmental matters drab and unglamorous, it has become increasingly difficult to brush the issue under the carpet, because the carpet itself is withering away.

The cartoons in the first chapter reflect upon man-made changes to the planet and their impact on the creatures we share our home with. Imagine Planet Earth on a global climate strike. The cartoons in this chapter are the cartoons you would find on her protest placards.

#2

ATLAS AND CLIMATE CHANGE

AH! SO YOU'RE THE BIGFOOT!
NAH. THAT'S JUST MY CARBON FOOTPRINT.

#3

THE GREAT BARRIER REEF ISN'T VISIBLE FROM SPACE ANYMORE, BUT HEY, THE GREAT PACIFIC GARBAGE PATCH IS!

#4

A GREEN TURTLE TAKES THE #BOTTLECAPCHALLENGE

WHACK!
KEEP YOUR TRASH OUT OF OCEANS.
-RWAN

#5

TONIGHT ON NATURE TV NEWS, MASS CORAL BLEACHING HITS THE MAJOR REEFS OF THE WORLD...
INCLUDING THE GREAT BARRIER REEF OF AUSTRALIA & THE LAKSHADWEEP REEF IN INDIA!
SCIENTISTS ASCERTAIN THE MAIN CAUSE TO BE AN UNPRECEDENTED RISE IN SEA SURFACE TEMPERATURES...
CHIEFLY ATTRIBUTED TO HUMAN-INDUCED CLIMATE CHANGE, GLOBAL WARMING, MINING, POLLUTION & OVER FISHING, AMONG OTHERS.
THE BLEACHING IS NOT ONLY CAUSING MASS DEATHS OF MARINE ORGANISMS, BUT IT IS ALSO AFFECTING HUMANS THEMSELVES WITH COASTAL COMMUNITIES ALREADY BEARING THE BRUNT OF ITS IMPACTS!
CORRESPONDENT CLOWNFISH JOINS US WITH LIVE UPDATES FROM THE REEFS...
WHAT DOES IT LOOK LIKE, MR. CLOWNFISH?
WELL, LET'S JUST SAY...
I KNOW A SPECIES THAT'S MAKING A REAL CLOWN OF ITSELF.
-RWAN

#6

IUCN* STATUSES OF SOME ANIMALS

(*INTERNATIONAL UNION FOR CONSERVATION OF NATURE)

CRITICALLY ENDANGERED
ENDANGERED
VULNERABLE
NEAR-THREATENED
LEAST CONCERN
LEAST CONCERNED

#7

NH-7...
NH-35...
NH-11...
80
NH-93...
MORE HIGHWAYS THROUGH THE NATIONAL PARK THAN STRIPES ON MY BODY!

THE APEX PREDATOR OF THE SEA STALKS ITS PREY...

#8

GOOD MORNING! I'M A MARINE BIOLOGIST AND I WELCOME YOU ALL TO THE 'MARINE ENVIRONMENT AWARENESS PROGRAMME'! ALL SET FOR YOUR FIRST ACTIVITY?
YESSS!
ALRIGHT! SO I'M GOING TO SHOVE PLASTIC WASTE DOWN YOUR THROATS, AND YOU'LL TELL ME HOW IT FEELS. READY?

#9

SOME COMMON SHOREBIRDS
THE AMERICAN GOLDEN PLOVER
THE RED KNOT
THE REDDISH EGRET
THE PECTORAL SANDPIPER
THE ROSEATE SPOONBILL
THE RUFF
SHOREBIRDS AFTER AN OIL SPILL
THE AMERICAN BLACK PLOVER
THE BLACK KNOT
THE BLACKISH EGRET
THE PETROLAR SANDPIPER
THE ROASTED SPOONBILL
THE ROUGHED UP

#10

3...2...1...
HAPPY NEW YEA—
BANG!
BOOM!
THERE GOES MY EAR.
-ROWAN

#11

WE WISH YOU A MERRY CONSUMERISTMAS
WE WISH YOU A MERRY CONSUMERISTMAS
WE WISH YOU A MERRY CONSUMERISTMAS
GIFTS a→z
GIFTS a→z
AND AN EAR-SPLITTING NEW YEAR!
BOOM
BANG
BANG
-ROWAN

#12

IT'S NOT MY CAR THAT'S POLLUTING THE AIR, IT'S THE BURNING OF CROP RESIDUE!
WHY SAY NO TO CRACKERS? CARS ARE POLLUTING THE AIR!
DON'T TELL ME TO USE PUBLIC TRANSPORT! I'VE WORKED HARD TO AFFORD THREE SUVs!
BAN CRACKERS, HUH? GO BAN BAKRI EID, YOU LIBTARD!
THE CITY WAS POLLUTED EVEN BEFORE I BOUGHT MY CAR!
CARS POLLUTED THIS CITY LONG BEFORE CRACKERS.
I'M AFRAID YOUR CHILD HAS BRONCHITIS.
I'M AFRAID YOUR CHILD HAS EMPHYSEMA AND HEARING DIFFICULTIES.
WHAT IS THE GOVERNMENT DOING TO CHECK POLLUTION?!
NOTHING!

#13

NARWHALS AND CLIMATE CHANGE

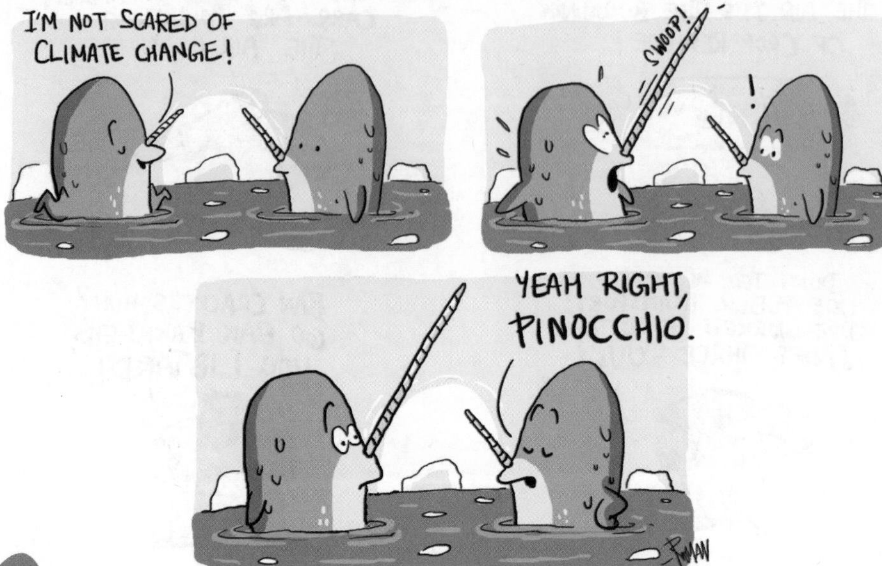
I'M NOT SCARED OF CLIMATE CHANGE!
SWOOP!
!
YEAH RIGHT, PINOCCHIO.

#14

WHAT THE HECK IS A PANDA DOING IN THE ARCTIC?!
I'M A POLAR BEAR WITH A TAN.

#15
OKAY, I ADMIT THAT I'M A MYTH.
NOW YOU'D BETTER ADMIT...
THAT CLIMATE CHANGE ISN'T.
NORTH POLE

#16
RUDOLPH THE RED-NOSED REINDEER HAS A LONG LIST OF WOES
DROWNING IN THE THAWED ARCTIC ICE HAS LEFT SANTA COMATOSE.
ALL OF THE OTHER REINDEER CAN'T FIND A SPECK OF LICHEN TO CONSUME.
THE WOLVES, THE FOXES & THE STARVING POLAR BEARS, ALL SEEM TO BE HEADED FOR DOOM.
ON ONE FOGGY CHRISTMAS EVE, THEY SET UP IN THE NORTH AN OIL RIG...
IT DUG AND DUG AND DUG AND DUG, UNTIL THERE WAS NOTHING TO DIG.
A SUDDEN SPARK BURST A CHAMBER OPEN, SPILLING AWAY ALL THE BLACK GOOEY OIL...
KILLING EVERY CREATURE IN ITS WAY FOR SEVERAL NAUTICAL MILES.
GLOBAL WARMING, CLIMATE CHANGE, AND THE OIL SPILLS CAUSED THE ARCTIC A HUGE SETBACK.
AND RUDOLPH'S FAMOUS SHINY RED NOSE IS NOW A LUMP OF BLACK.

#17

OFFBEAT OLIVIA'S OFFBEAT TRAVELS!
FLIES TO A NEW COUNTRY EVERY WEEK.
TAKES SELFIES WITH CAPTIVE WILD ANIMALS
TIGER TEMPLE
ENDORSES CIVET COFFEE
KOPI LUWAK
ENDORSES EXOTIC/ GAME MEAT. (SO OFFBEAT!)
RELISHES DURIAN, PALM OIL PRODUCTS AND ALL MONOCULTURE PRODUCE.
NEW OUTFITS & ACCESSORIES IN EVERY PICTURE.
SHOPS COMPULSIVELY FOR MEMORABILIA.
HASHTAGS 'SUSTAINABLE TOURISM' IN ALL HER POSTS.
ECO CHICK
RAJAN

#18

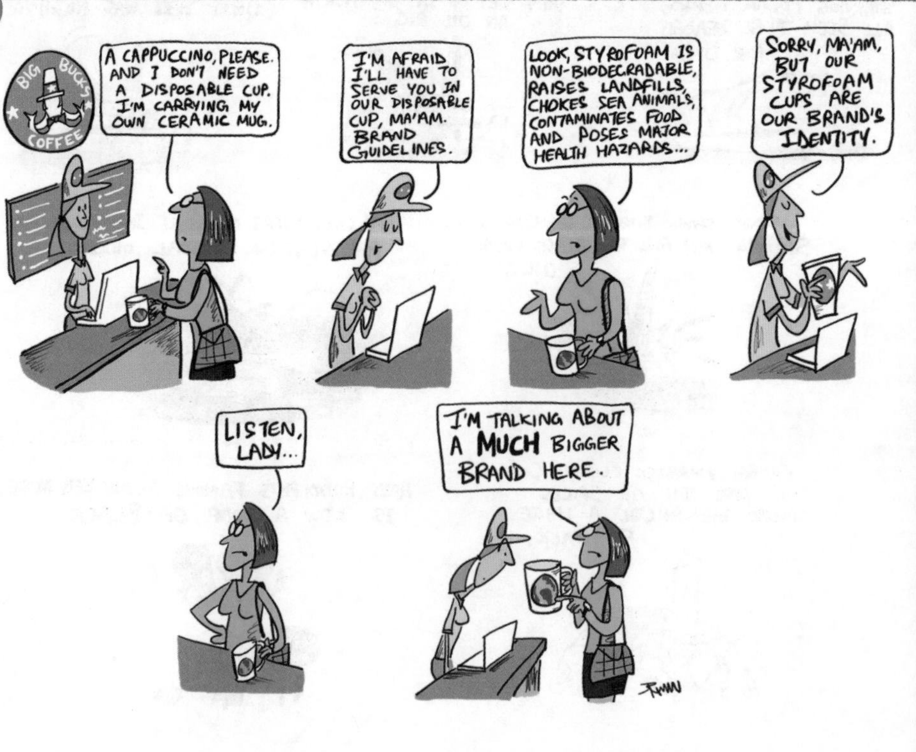
BIG BUCKS COFFEE
A CAPPUCCINO, PLEASE. AND I DON'T NEED A DISPOSABLE CUP. I'M CARRYING MY OWN CERAMIC MUG.
I'M AFRAID I'LL HAVE TO SERVE YOU IN OUR DISPOSABLE CUP, MA'AM. BRAND GUIDELINES.
LOOK, STYROFOAM IS NON-BIODEGRADABLE, RAISES LANDFILLS, CHOKES SEA ANIMALS, CONTAMINATES FOOD AND POSES MAJOR HEALTH HAZARDS...
SORRY, MA'AM, BUT OUR STYROFOAM CUPS ARE OUR BRAND'S IDENTITY.
LISTEN, LADY...
I'M TALKING ABOUT A MUCH BIGGER BRAND HERE.
RAJAN

#19
THE BIG, FAT INDIAN WEDDING'S BIGGER, FATTER CARBON FOOTPRINT
FLIGHTS FOR 1800 GUESTS
3600 MULTIPAGE PRINTED INVITES
500 PRIVATE SEDANS DRIVING GUESTS AROUND THE CITY
A NIGHT-LONG SHOW OF AIR & NOISE-POLLUTING FIRE CRACKERS
NOISE-POLLUTING WOOFERS BLARING GARISH BOLLYWOOD MUSIC ALL NIGHT
POINTLESS, NIGHT-LONG USE OF WEDDING LIGHTS
PURCHASE OF EXORBITANT APPAREL & JEWELLERY, NEVER TO BE WORN AGAIN
TONS OF FOOD & FLORAL WASTE
TONS OF PLASTIC & STYROFOAM WASTE
I DON'T SEE A BRIDE & GROOM. I SEE TWO GIANT LANDFILLS GETTING MARRIED.
-RMAN

#20
RECENT SIGHTINGS FROM MOUNT EVEREST
COLOURFUL PRAYER FLAGS: BASE CAMP
EVEN MORE COLOURFUL PLASTIC LITTER: KHUMBU ICEFALL
DISCARDED TENTS AND CLIMBING GEAR: CAMP 1
HUMAN SEWAGE: CAMP 2
UGH!
MORE PLASTIC WASTE: CAMP 3
TRASH-PICKING YETI: SUMMIT

#21
THIS IS THE PANGONG TSO, A LAKE IN LADAKH CLASSIFIED AS A 'RAMSAR WETLAND OF GLOBAL IMPORTANCE'.
EVER SINCE THE HINDI FILM 3 IDIOTS WAS SHOT HERE, THE LAKE HAS BECOME A HUB FOR TOURISTS, AND OF COURSE, PLASTIC LITTER.
COUNTLESS RESORTS & HOTELS HAVE SPRUNG UP ACROSS LADAKH, PUTTING TREMENDOUS PRESSURE ON ITS LIMITED GROUNDWATER AND NATURAL RESOURCES...
AND CRACK-BRAINED ACTIVITIES LIKE OFF-ROADING RALLIES ENDANGER ITS FRAGILE ECOSYSTEM.
WHILE TOURISM DOES HAVE THE POTENTIAL TO UPLIFT AND EMPOWER LOCAL ECONOMY, UNSUSTAINABLE NUMBERS AND METHODS ARE A RISING THREAT TO LADAKH.
I'VE LOST COUNT OF HOW MANY IDIOTS HAVE SHOWED UP SINCE THE FIRST THREE.

#22

LADIES, GENTLEMEN, GIRLS & BOYS! WELCOME TO YOUR FAVOURITE AIRLINE!
WE SHALL SHORTLY BEGIN SERVING YOU BREAKFAST & REFRESHMENTS FROM OUR DELECTABLE IN-FLIGHT MENU
ASSORTED FRUITS IN A STYROFOAM BOX WRAPPED IN CELLOPHANE
DISPOSABLE SUGAR & CREAM SACHETS
BOTTLED MINERAL WATER. (MORE PLASTIC BOTTLES FOR SECOND HELPINGS, NOT REFILLS.)
PLASTIC STIRRER
COFFEE IN A DISPOSABLE CUP. (MORE CUPS FOR SECOND HELPINGS, NOT REFILLS.)
MILK ON THE SIDE IN A FANCIER-LOOKING PLASTIC CUP
BUTTER IN A DISPOSABLE PLASTIC CONTAINER
BREAD WRAPPED IN CELLOPHANE
AND PLASTIC CUTLERY, OF COURSE
JAM IN A DISPOSABLE PLASTIC CONTAINER
SANDWICH WRAPPED IN CELLOPHANE
OMELETTE WRAPPED IN CELLOPHANE
SALAD IN A PLASTIC BOX WRAPPED IN CELLOPHANE
THE ENTIRE ARRANGEMENT IN YET ANOTHER STYROFOAM CONTAINER WITH A CELLOPHANE COVER
WE HOPE YOU ENJOYED YOUR MEAL! WOULD YOU LIKE TO PURCHASE OUR 'GREEN FLYER' CARBON OFFSETS?
-Rohan

#23

WHY DO GOLFERS DO THAT?
ELITISM-INDUCED MYOPIA FROM PLAYING ONE OF THE WORLD'S MOST ENVIRONMENTALLY UNSUSTAINABLE SPORTS.
-Rohan

#24

KIRIBATI: AMONG THE FIRST NATIONS EXPECTED TO SUBMERGE BECAUSE OF CLIMATE CHANGE.

WHOEVER DESIGNED THIS FLAG HAD GREAT FORESIGHT.

#25

GEOGRAPHICALLY, THE ANDAMAN & NICOBAR ISLANDS MAKE UP JUST 0.25% OF INDIA'S LAND MASS. BUT GUESS HOW MUCH OF INDIA'S FAUNA THE ISLANDS HOST?
10%!
AND CAN YOU GUESS HOW MANY OF THESE SPECIES ARE FOUND EXCLUSIVELY ON THESE ISLANDS?
OVER 1050 SPECIES. YOU HEARD THAT RIGHT!
THIS STAGGERING ENDEMISM IS A RESULT OF MILLIONS OF YEARS OF GEOGRAPHIC ISOLATION.
BUT POACHING, LOGGING, URBANIZATION & PRESSURES FR TOURISM ARE FAST DISPLACING US ENDEMICS FROM OUR VERY LAND!
IN THIS RUSH TO 'DEVELOP' THE ANDAMAN AND NICOBAR ISLANDS, LET'S NOT FORGET THE BOTTOMLINE-
PROTECTING ENDEMISM IS PROTECTING 'ANDAMISM'!
-ROHAN

#26
SOME ENVIRONMENTAL IMPACTS OF AN INDO-PAK WAR
INCALCULABLE DESTRUCTION OF A FRAGILE HIMALAYAN ECOSYSTEM
IRREVERSIBLE DAMAGE TO GLACIERS- THE ORIGINS OF OUR FRESH WATER
RENDERING WILDLIFE-MONITORING IMPOSSIBLE, GIVING POACHING A FREE HAND
RELEASE OF TOXIC CHEMICALS INTO THE VERY SOURCES OF OUR RIVERS
GREENHOUSE FOOTPRINTS OF BOTH WARRING NATIONS HITTING NEW HIGHS...
EVEN AS MANKIND STOOPS TO NEW LOWS.

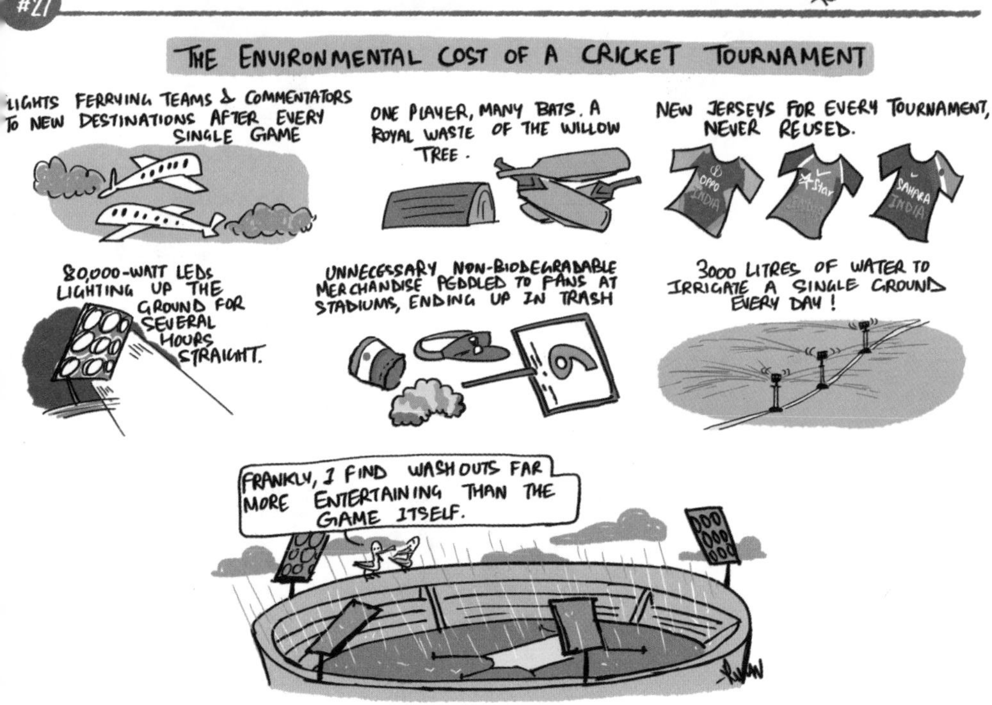
#27
THE ENVIRONMENTAL COST OF A CRICKET TOURNAMENT
LIGHTS FERRYING TEAMS & COMMENTATORS TO NEW DESTINATIONS AFTER EVERY SINGLE GAME
ONE PLAYER, MANY BATS. A ROYAL WASTE OF THE WILLOW TREE.
NEW JERSEYS FOR EVERY TOURNAMENT, NEVER REUSED.
OPPO INDIA
Star
SAHARA INDIA
80,000-WATT LEDS LIGHTING UP THE GROUND FOR SEVERAL HOURS STRAIGHT.
UNNECESSARY NON-BIODEGRADABLE MERCHANDISE PEDDLED TO FANS AT STADIUMS, ENDING UP IN TRASH
3000 LITRES OF WATER TO IRRIGATE A SINGLE GROUND EVERY DAY!
FRANKLY, I FIND WASHOUTS FAR MORE ENTERTAINING THAN THE GAME ITSELF.

#28

LET US NOW TRY TO UNDERSTAND HOW E-WASTE HAS BECOME ONE OF THE MOST RAPIDLY EXPANDING MUNICIPAL WASTE STREAMS.
PROJECTOR ERROR
DUMP THIS PIECE OF TRASH AND GET ME A NEW HD PROJECTOR RIGHT NOW

#29

OKAY, LET'S GO THROUGH THE LIST OF ESSENTIALS FOR OUR EVENT ONCE AGAIN.
NOTEPADS & DISPOSABLE PLASTIC PENS FOR THE AUDIENCE?
CHECK!
BOTTLED WATER FOR ALL ATTENDEES?
CHECK!
AIR CONDITIONING, HIGH-DEFINITION PROJECTION SCREENS, STEREO SPEAKERS & ELECTRONIC ACCESSORIES?
CHECK!
GLASS MEMENTOS FOR ALL DIGNITARIES, GIFT-WRAPPED IN CELLOPHANE?
CHECK!
TWO-WAY AIR TICKETS FOR ALL INVITEES?
CHECK!
1500 PRINTED INVITES?
CHECK!
PERFECT! THIS SUSTAINABILITY SUMMIT WILL BE A MEGA SUCCESS!

#30

THE ARCTIC WILDFIRES OF 2020

HEY, OIL COMPANIES, FOSSIL FUEL INDUSTRIES, PRO-FOSSIL FUEL GOVERNMENTS AND CLIMATE CHANGE DENIERS OF THE WORLD...
ARE YOU NOT ENTERTAINED?!
THE SIBERIAN ARCTIC

#31

WHAT'S THIS?
IT'S CALLED A STRAW.
IT'S A HOLLOW, PLASTIC SUCKING DEVICE INVENTED BY MAN.
TO SUCK WHAT?
SEVERAL THINGS.
SUCKING DRINKS FROM A BOTTLE OR A GLASS...
SUCKING THE LIFE OUT OF THIS PLANET...
AND SUCKING AT LIFE, IN GENERAL.
WOW. THE ONLY PARASITE THAT SUCKS USING A DEVICE. AN EVOLUTIONARY MARVEL!
INDEED.

#32

LOOKING AT ALL THIS PRISTINE GREENERY, THIS BOUNTIFUL FRONDESCENCE, REALLY MAKES YOU WONDER IF CLIMATE CHANGE & THE ANTHROPOCENE ARE JUST MEDIA HOAXES & YOUTH DELUSIONS...
UNTIL YOU REALIZE YOU'RE FLYING OVER ANTARCTICA.

#33

NOTHING ROYAL ABOUT THIS.
KING-LIKE NEITHER.
EMPEROR MY @**.

THE ROYAL, KING AND EMPEROR PENGUINS DISCUSS CLIMATE CHANGE.

Dear Ma'am,
it has been my honour being at your service all these years...
But it is with deep regret that I write to you today, on this eve of Diwali.
As you are well aware, my acute hearing helps me hunt rats in the dark—a skill that has enabled me to render free pest-control services to mankind for thousands of years.
But tonight, the sounds of thousands of exploding crackers will potentially deafen me, disabling my hunting abilities forever.
The air will be rendered unbreathable from all the pollution.
And to make it worse, thousands of us will be captured illegally to be offered as religious sacrifices!
Ma'am, the owls of this country deserve a LOT more respect.
And thus, unless you, in your capacity as Honourable Goddess of Wealth, have ensured that crackers and owl-poaching have been banned in practice throughout India,
I, your 'Vahana', announce my resignation from your office.
Wishing you a happy Diwali,
Yours faithfully, Barn Owl
sniff
NOW TO MAKE IT TO THE POSTBOX ALIVE!

#35

RINNGGG!
YAWN.
IS IT MY HANGOVER OR ARE WE REALLY ON MARS?!
NEITHER, HONEY. IT'S A GLOBAL-WARMING INDUCED ALGAL BLOOM IN ANTARCTICA.
AND MANKIND'S ALARM CLOCK HAS STILL NOT RUNG.

#36

COMBAT CLIMATE CHANGE. STOP POPULATION GROWTH WORLDWIDE.
FLAWED!
BLAME BIG CORPORATIONS, NOT PEOPLE!
MALTHUSIAN!
PRIVILEGED @#*%!
HOW DARE YOU?!
HAVE REPRODUCTIVE SYSTEM, WILL REPRODUCE!
MM.OKAY.
BIG CORPORATIONS, STOP POLLUTING.
HUH?! HAVE GOVERNMENT SUPPORT, WILL POLLUTE!
HMM.
GOVERNMENTS, STOP BIG CORPORATIONS FROM POLLUTING.
BUT JOBS?
EMPLOYMENT?
SUSTAINING GROWING POPULATIONS?
HAVE CORPORATION MONEY, CAN'T.
WELL, I GUESS I'LL JUST HAVE TO WAIT FOR YOU GUYS TO COME UP WITH A SOLUTION.

#37

WRAPPED™
YOUR ONE STOP DAILY SHOP
WHAT IS THE NEED FOR ALL THIS PLASTIC?! I NEED TO SPEAK TO THE OWNER.
ERR... THIS WAY, MA'AM.
AT A TIME WHEN THE WORLD IS WAKING UP TO THE PERILS OF PLASTIC POLLUTION, YOUR STORE CONTINUES TO SELL FRUITS & VEGGIES WRAPPED IN PLASTIC! THIS IS RIDICULOUS!
MISTER, I'M TALKING TO YOU.
PROCESSING...
ERROR!

#38

FRIENDS! I, CAPITALISM, FINALLY ADMIT THAT I CREATED THE AIR CRISIS.
BUT NOW I PRESENT TO YOU THE VERY CURE FOR IT—
THE AIR PURIFIER™!
IS CLEAN AIR ONLY FOR THE PRIVILEGED?! WHAT ABOUT THE POOR WHO BEAR THE BIGGEST BRUNT OF IT?!
NOW GET DESIGNER POLLUTION MASKS™ FREE WITH EVERY AIR PURIFIER™!
YAYYY!

#39

WILDLIFE PHOTOGRAPHY EXHIBITIONS THEN
WILDLIFE PHOTOGRAPHY EXHIBITIONS NOW

#40

CLIMATE CHANGE...
RETREATING GLACIERS...
LITTERED MOUNTAINS...
HABITAT LOSS AND RAPIDLY DETERIORATING MOUNTAIN ECOLOGY...
ASK ME ONE MORE TIME WHY PALLAS'S CATS ALWAYS LOOK GRUMPY.

#41

CONGRATULATIONS ON BEING GRANTED THE SAME LEGAL RIGHTS AS PEOPLE, RIVER GANGA!
THANKS, GANGETIC DOLPHIN. BUT I DON'T SEE HOW IT MAKES ANY SENSE, WHEN MY LEGAL RIGHTS ARE VIOLATED EVERY DAY.
AS WE SPEAK, SAND MINING, DUMPING OF SEWAGE, GARBAGE & INDUSTRIAL WASTE, AND ILLEGAL FISHING GO ON UNABATED...
I'M LIVING ON THE VERY EDGE, AND ONE SMALL MISTAKE COULD SPELL DOOM!
UH OH! HERE COME THE OFFERINGS FROM YOUR 'DEVOTEES' IN PLASTIC BAGS!
ACK!
COUGH!
CHOKE
HELP!...
IT MAY BE A BIT LATE TO GRANT HER LEGAL RIGHTS...
SOMEBODY PLEASE CALL AN AMBULANCE!
RMAN

TWO

Man–Animal Conflict

TIGER NUMBERS RISE!! SAY YAY!!
TBN THE BIG INDIAN NEWS
YAYYY.

#1

Whoever came up with the term 'Man-Animal Conflict' was a diplomat, desperate to sound politically correct, because in this equation 'conflict' can be attributed to our species. But don't you worry, the cartoons in the chapter are anything but politically correct.

From trawling in the sea to hunting birds in the air, from trophy hunting in Africa to poaching in China, from the impacts of palm oil in South East Asia to the illegal pet trade in the Amazon Basin, these cartoons take a dig at the many innovative ways man has come up with to wreck the planet we call home.

#2

THE GOOD NEWS -
CUSTOMS OFFICIALS SEIZE RHINO HORNS & IVORY; BUST ANOTHER SMUGGLING RACKET.
THE BAD NEWS -
THOSE RHINOS & ELEPHANTS AREN'T CELEBRATING.
-RWAN

#3

THIS IS A RHINO HORN.
THIS IS HUMAN HAIR.
AND THIS IS A HUMAN FINGERNAIL.
GUESS WHAT'S COMMON AMONG THE THREE?
THEY'RE THE SAME DAR
SUBSTANCE - KERATIN!
SO, THE NEXT TIME YOU'RE IN THE MOOD FOR SOME PHONEY VIETNAMESE APHRODISIAC (WHICH YOUR SPECIES REALLY DOESN'T NEED, GOING BY THE RATE AT WHICH YOU GUYS REPRODUCE...)
PLEASE JUST CHEW ON YOUR DARN NAILS OR YOUR HAIR INSTEAD OF BRUTALLY POACHING AN ENDANGERED ANIMAL THAT'S STRUGGLING TO SURVIVE AND REALLY DOESN'T WANT ANYTHING TO DO WITH YOU!
OKAY, THEN. TRUCE?
BANG!
SLASH!
ONE THING'S FOR SURE.
NO RHINO COULD BEAT YOU GUYS AT BEING THICK-SKINNED
-RWAN

#4

LOOKS LIKE WE'VE DISCOVERED AT LEAST ONE GOOD USE OF PALM OIL.
PALM OIL
PALM OIL CO.

5

THIS IS 'KOPI LUWAK,' THE WORLD'S MOST EXPENSIVE COFFEE...
KOPI LUWAK
AND I AM ITS SOURCE, AN ASIAN PALM CIVET.
CIVET COFFEE IS MADE FROM UNDIGESTED COFFEE BEANS IN MY DROPPINGS.
THE POPULARITY OF CIVET COFFEE HAS LED SMALL-SCALE COTTAGE INDUSTRIES ACROSS SOUTH EAST ASIA TO GROW INTO INTENSIVE FARMS...
WHERE HUNDREDS OF WILD CIVETS ARE CAPTURED IN CRAMPED, FILTHY CAGES, AND FORCE-FED COFFEE BEANS TO INCREASE YIELDS.
A WILD, THREATENED ANIMAL HELD CAPTIVE AND BRUTALIZED FOR A BEVERAGE!
A CUTE, CUDDLY CREATURE ABUSED FOR A CUP OF COFFEE!
AN OMNIVORE DENIED ITS NATURAL DIET, & FORCE-FED! IS THIS ETHICAL? IS THIS JUST? IS THIS IN ACCORDANCE WITH CONSERVATION?
SIGH.
I COULD REALLY USE AN ESPRESSO RIGHT NOW.

#6

OKAY, I HAVE A RIDDLE FOR YOU— WHERE CAN YOU FIND A TIGER & A PANGOLIN IN CLOSEST PROXIMITY?
HMM... AN ANT HILL? ...A BURROW? ...A COMIC STRIP?
NO. NO. NO.
GASP!
A BOTTLE OF CHINESE MEDICINE.
TCM

#7

THE BEAR BILE RHYME

BAA, BAA, BLACK BEAR, HAVE YOU ANY BILE?
YES, BUT NOT FOR YOU, YOU IMBECILE!
IT DOESN'T CURE COVID IT DOESN'T CURE S#%T!
LET ME OUT OF THIS CRAMPED CAGE, YOU ****ING DIMWIT!

#8

CLICK
INSTANT TROPHY PICTURES!
FOR ENQUIRIES, CALL
IT'S A FREE SERVICE WE'VE STARTED TO BRING DOWN TROPHY HUNTING.

9

HEY, RANGER. YOU SHOT THE AFRICAN POACHER THE OTHER DAY. WHY AREN'T YOU SHOOTING THIS WHITE POACHER?
HE'S NOT A POACHER! HE'S A TROPHY HUNTER. HE PAYS A HEFTY FEE TO THE GOVERNMENT AND HELPS CONSERVATION.
OH. HOW EXACTLY? BY TAKING ANOTHER ANIMAL OUT OF AN ALREADY PLUMMETTING POPULATION?
BY IMPORTING HIS 'TROPHY' BACK TO HIS COUNTRY, FURTHER FUELLING THE DEMAND FOR IVORY?
OR IS HE CALLED A TROPHY HUNTER, AND NOT A POACHER, SIMPLY BECAUSE OF HIS SKIN COLOUR & PRIVILEGE?
AW! LOOK AT YOU DOCUMENTING THIS HISTORIC MOMENT IN ELEPHANT 'CONSERVATION'.
CLICK

#10

HAVING SCHEDULE 1 PROTECTION UNDER THE WILDLIFE PROTECTION ACT FEELS SO SPECIAL...
ON PAPER.

#11

YOU KNOW WHAT THE PROBLEM WITH LONGLINE FISHING IS?
WHAT?
THIS LOOOOOONG LINE OF DEAD, ENDANGERED MARINE WILDLIFE.

12

HERE'S THE BUTTER GARLIC SHRIMP YOU ORDERED, SIR.
AH, THANK YOU.
AND HERE'S THE BYCATCH TRAWLED ALONGWITH THE EIGHT PIECES OF SHRIMP ON YOUR PLATE...
THE SELECTION INCLUDES SHARKS, RAYS, COD, SEAHORSES, STARFISH, REEF FISH, AND EVEN TURTLES, SCOOPED AWAY INDISCRIMINATELY, ONLY TO BE DISCARDED.
ENJOY YOUR MEAL, SIR. AND DON'T WORRY; THE BYCATCH IS COMPLIMENTARY.

13

HILSA, OR 'ILISH' AS THE BENGALIS CALL IT, IS THE PRIDE OF BENGALI CUISINE.
BUT ILISH STOCKS HAVE BEEN CRASHING ALARMINGLY OVER THE YEARS.
ILISH'S PROBLEMS BEGAN WHEN THE FARAKKA BARRAGE, BUILT IN 1975, PERMANENTLY BLOCKED THE ANADROMOUS FISH'S UPSTREAM MIGRATION.
TODAY, ILISH STOCKS ARE THREATENED BY GILLNETS & SEINE NETS, THAT DO NOT SPARE EVEN JUVENILE FISH...
WHILE MECHANIZED TRAWLERS CONTINUE TO DEPLETE FISH STOCKS, DESTROY HABITATS, AND THREATEN ARTISANAL FISHING.
THE EVER-INCREASING DEMAND FROM CONSUMERS, SPECIALLY IN THE FESTIVE SEASON, ONLY MAKES MATTERS WORSE...
PUSHING FISHERMEN TO FLOUT SEASONAL BANS, LEAVING ILISH POPULATIONS WITH NO SCOPE FOR RECOVERY.
NO WONDER I'VE BEEN FEELING SO ILL-ISH LATELY!

#14

THE WORLD'S SMALLEST MONKEY, THE PYGMY MARMOSET OF PERU...
IS CAPTURED FROM ITS RAINFOREST HOME...
SNATCHED FROM ITS PARENT...
TRANSPORTED IN A CRAMPED CAGE...
AND SOLD TO EXOTIC PET OWNERS AS THE 'FINGER MONKEY'!

#15

HERE'S A LIST OF CONSERVATION THREATS FACED BY SMOOTH-COATED OTTERS:
1. Loss of habitat to dams.
2. Destruction of wetlands for agriculture.
3. Poaching for fur.
4. Water pollution.
5. Pesticidal contamination.
6. Over fishing & depletion of prey base.
MIND YOU, BEING A SMOOTH-COATE OTTER IS N SMOOTH RIDE

I'M THE GREY SLENDER LORIS, A NOCTURNAL PRIMATE FOUND ONLY IN SOUTH INDIA & SRI LANKA...
BECAUSE I'M ADAPTED FOR AN ARBOREAL LIFE, I NEED CONTINUOUS STRETCHES OF FORESTS TO MOVE ABOUT...
I GUESS THE 'SLENDER IN MY NAME REFERS TO CHANCES OF SURVIVAL

THIS IS A PET.
THIS IS A PET.
THIS IS ALSO A PET.
THIS IS NOT A PET!
I'M A WILD ANIMAL, I LIVE IN THE RAIN-FORESTS, AND THE EXOTIC PET TRADE IS THREATENING MY SPECIES.
I'M NOT A DOG OR A CAT OR A @#%*ING STUFFED BEAR! I'M A SLOW LORIS. GET IT?
YOU GUYS ARE SLOW, AREN'T YOU?
ROMAN

#17

EMBARRASSMENT.
THAT'S WHAT STOPPED INDIA FROM MAKING ME, THE GREAT INDIAN BUSTARD, ITS NATIONAL BIRD. THE FEAR OF EMBARRASSMENT FROM BEING MISSPELLED AS 'GREAT INDIAN B*STARD.'
I'LL TELL YOU WHAT EMBARRASS INDI A LOT MORE TH A SPELLIN MISTA
THAT A ONCE-NATIONAL BIRD CONTENDER, AND THE STATE BIRD OF RAJASTHAN, IS ON THE VERY BRINK OF EXTINCTION!
THAT DESPITE KNOWING WHAT KILLS US— HABITAT LOSS, COLLISIONS WITH OVERHEAD TRANSMISSION LINES & WINDMILLS— NOTHING HAS BEEN DONE!
AND THAT THE GREAT INDIAN BUSTARD IS ALL SET TO BE THE FIRST BIRD TO GO EXTINCT IN INDEPENDENT INDIA!
NOW WHERE ARE ALL THOSE BELLOWING NEWS ANCHORS? THOSE HOWLING DEBATES? THOSE AGITATED MOBS? THOSE VENOM-SPEWING TWEETS?
YOUR PSEUDO PATRIOTISM EMBARRASSE ME.

#18

Tracing Shark Fin Soup Back to Its Source

One shark fin in a steaming soup bowl in a high-end Oriental restaurant
A shipment of tons of shark fin on a seafood transport cargo
A live finned shark being dumped off a shark finning vessel along a tropical coast
A tropical coral reef being obliterated by reef-eating fish in the absence of an apex predator
Disappearance of reefs & fish from the region
The collapse of a fishing industry
? ? ?
When a finned shark drowns, it never drowns alone.
Wink

#19

I NEVER THOUGHT I'D SAY THIS, BUT I ACTUALLY PREFER BEING TROLLED ON TWITTER THAN BEING TRAWLED.
HEY, WAIT A SECOND! I DON'T REMEMBER SIGNING ANY CLEARANCE PAPERS!
PARK
ROAD WORK AHEAD

#20

eshop
REAL Tiger Claw $1000
Instascam
PET OTTER FOR SALE! $2000
facethat
ELEPHANT IVORY TRINKET: $6000
RHINO HORN $10000
Google
how to sue the internet
Tap
Tap
Tap

#21

MITTHU, MY LITTLE PARROT, SPEAK!
AH, SO YOU WANT ME TO SPEAK! OKAY, TWO THINGS...
FIRST, I'M NO 'LITTLE PARROT.' I'M AN ALEXANDRINE PARAKEET, ONE OF THE LARGEST WILD PARROTS FOUND IN INDIA.
AND SECOND, BECAUSE OF GUYS LIKE YOU, WILD PARROTS FACE A SERIOUS THREAT FROM THE ILLEGAL PET TRADE...
WHICH, BY THE WAY, IS PUNISHABLE WITH FINE & IMPRISONMENT UNDER THE WILDLIFE PROTECTION ACT.
SO IDEALLY, I SHOULD BE BACK WITH MY FOLKS IN MY DECIDUOUS FOREST HOME, AND YOU SHOULD BE SERVING A JAIL SENTENCE INSTEAD OF ME.
ALRIGHT. I'M DONE SPEAKING. NOW YOU SPEAK.
SPEAK, BUDDY. THE KID IS WATCHING.

#22

HOW TO MAKE ART WITHOUT KILLING A MONGOOSE
– A guide for artists by the Grey Mongoose.
SAY NO TO MONGOOSE HAIR BRUSHES
LEARN TO IDENTIFY MONGOOSE HAIR BRUSHES
Stiff hair pointing steadily upwards
Colour gradient from dark brown at roots & tips to cream in the centre
SPREAD THE WORD TO YOUR FELLOW ARTISTS & FANS
MY ART IS #mongoose hair free
APPRECIATE OUR CONTRIBUTION: SAVING MILLIONS WORTH OF CROPS EACH YEAR BY PREYING ON PESTS.
CONSIDER SWITCHING TO THE LEAST HARMFUL MEANS OF CREATING ART AVAILABLE TO YOU
EDWARD'S DIGIPAINT TUTORIALS
SUBSCRIBE NOW!
LEARN MORE ABOUT THE SIX SPECIES OF MONGOOSES IN INDIA
GREY
RUDDY
SMALL ASIAN
BROWN
STRIPE-NECKED
CRAB-EATING
CELEBRATE MONGOOSES IN YOUR ART
DRAW ME LIKE ONE OF YOUR WILD GIRLS!

#23

NUMBERING JUST 7000, CHEETAHS ARE NOW 'VULNERABLE' ON THE IUCN RED LIST.
HABITAT LOSS HAS CONFINED US TO LESS THAN 10% OF OUR FORMER RANGE.
POACHING FOR USE OF OUR PARTS IN CHINESE MEDICINE CONTINUES TO THREATEN OUR POPULATION...
AND TO MAKE IT WORSE THAN EVER OUR CUBS ARE NOW SMUGGLED TO THE GULF AS PETS- AN ILLEGAL TRADE FUELLED BY SOCIAL MEDIA!
YOU KNOW, GAZELLE, I CAN NEVER TELL IF I'M RACING WITH YOU OR JUST RACING AWAY TO EXTINCTION.

#24

TIGER PARKS THEN
TIGER PARKS NOW
LUXURY TIGER RESORT
TIGERS & TIGERS LUXURY RESORT & SPA
ANOTHER TIGER RESORT
THE TIGER MEADOWS INN
THE NEW TIGER RESORT

#25

WELCOME BACK TO NATURE TV! DID YOU KNOW THAT THE GOBI BEAR IS THE WORLD'S ONLY DESERT BEAR?
AND THE MOST DESERTED ONE TOO...
WITH JUST 22 REMAINING, WE'RE ALMOST EXTINCT!
THANKS TO CLIMATE CHANGE & EXTENDED DROUGHTS, THE VEGETATION WE THRIVE ON IS GETTING SCARCER...
PLUS THERE'S COMPETITION WITH THE EVER-GROWING LIVESTOCK & THE THREAT FROM POACHING!
THE GOBI BEARS NEED YOUR SUPPORT. WE NEED URGENT PROTECTION. WE NEED A VOICE.
I SHOULD BE HOSTING THIS SHOW!

#26

10... 11... FASTER! 12...
13... 14... HARDER! 15...
16... 17... GOOD! 18...
YEEAOW!
NOW YOU'RE READY, MY BOY!
UNTIL INDIA GETS SERIOUS ABOUT MITIGATION, ALL YOU HAVE IS BRACHIATION.
NATIONAL PARK
NH7

#27

#dolphinselfie
#lookatmeladies
#snakeselfie
#fearless
#totalStud
#octopusselfie
#seadude
#nofear
#bearselfie
#imbeargrylls
#YOLO
WHACK!
#wildanimalsarent selfieprops
#leaveusalone
#YOLO indeed

#28

GOODNESS! A HUNDRED MILLION BIRDS ARE ESTIMATED TO BE KILLED ANNUALLY BECAUSE OF COLLISIONS WITH WINDOWS!
MOVE TO MAC, DUDE.

#29

AREN'T YOU THE INDO-PACIFIC HUMPBACKED DOLPHIN?
YES!
I'VE BEEN MEANING TO ASK YOU A QUESTION...
SURE.
DESPITE ALL THE THREATS YOU FACE — DESTRUCTION OF COASTS BY PORTS, WATER POLLUTION, NOISE FROM SHIPS, COLLISIONS WITH BOATS, OVERFISHING, AND EVEN POLLUTANTS CONTAMINATING YOUR MILK, KILLING YOUR BABIES...
MMHMM...
WHY ARE YOU ALWAYS SMILING?!
I SMILE BECAUSE I KNOW THAT WHAT MANKIND HAS DONE TO THE PLANET, THE PLANET WILL DO TO IT SOONER OR LATER.
REVENGE IS THE FUNNIEST PRACTICAL JOKE.

#30

YANKEE DOODLE: FEAT. THE YANGTZE PADDLEFISH

YANGTZE PADDLE WENT TO TOWN
A-RIDING IN A FISH NET
DAMS HAD WIPED HIS SPECIES OUT
BEFORE THE WORLD COULD RE-GRET.
La la la la la la la...
Psephurus gladius
EXTINCT!

#31

CLICK CLICK CLICK
CLICK CLICK CLICK
See-see-tee-twooti twooti
WAIT A SECOND. WHEN WAS TROPHY HUNTING LEGALIZED IN INDIA?!
CLICK CLICK CLICK

#32

WE'RE THE SAIGA ANTELOPE OF THE CENTRAL ASIAN STEPPES.
FIRST WE WERE WIPED OUT OF ROMANIA & MOLDOVA BY HUNTERS...
THEN POACHED TO EXTINCTION IN CHINA FOR OUR HORNS, A PRIZED INGREDIENT IN CHINESE MEDICINE...
HUNDREDS OF US ARE STILL KILLED ANNUALLY FOR OUR MEAT...
TO ADD TO THAT, A MASS DIE-OFF IN AN EPIZOOTIC OUTBREAK IN 2015 KILLED HALF OUR POPULATION...
AND YET ANOTHER EPISODE OF MASS DEATHS HAPPENED IN 2016, FROM GOAT PLAGUE!
SAIGAS NEED YOUR HELP MORE THAN EVER...
IT'S BEEN AN EPIC SAIGA OF SURVIVAL STRUGGLES!

#33

STOP! BEFORE YOU SHOOT, YOU MUST SIGN THIS AGREEMENT.
In the absence of Fruit Bats, I, Mr./Mrs. _______ hereby agree to:
1. Conduct long-range Pollination & seed dispersal of native flora.
2. Ensure continual long-distance gene flow between various fruit trees.
3. Maintain the general good health of the eco-system.
4. Re-check the meaning of 'pest' in the dictionary.
(signature)

#34

21st CENTURY PROBLEMS:

JUST 21 LIKES ON MY FACEBOOK POST!
JUST 21 RETWEETS FOR MY TWEET!
JUST 21 HITS ON MY YOUTUBE VIDEO!
JUST 21 MEMBERS OF MY SPECIES ALIVE.
VAQUITA REFUGE AREA

HOW DOES MY NEW CORAL JEWELLERY LOOK ON ME, SWEETIE?
HMM. LET'S SEE. NO ASSOCIATED ZOOXANTHELLAE. NO NATURAL PIGMENTATION. NO CLEANING STATIONS, HABITATS OR SPAWNING GROUNDS FOR REEF FISH.
FRANKLY, A LITTLE AWARENESS AND COMMON SENSE WOULD LOOK LIKE MUCH BETTER ACCESSORIES ON YOU, MUM.

#35

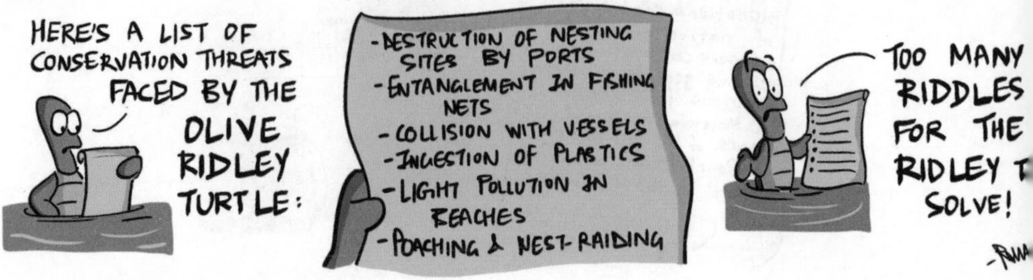

HERE'S A LIST OF CONSERVATION THREATS FACED BY THE OLIVE RIDLEY TURTLE:
-DESTRUCTION OF NESTING SITES BY PORTS
-ENTANGLEMENT IN FISHING NETS
-COLLISION WITH VESSELS
-INGESTION OF PLASTICS
-LIGHT POLLUTION IN BEACHES
-POACHING & NEST-RAIDING
TOO MANY RIDDLES FOR THE RIDLEY T
SOLVE!

#36
WHAT PEOPLE THINK THEY'RE DOING BY FEEDING MONKEYS
PAYING RELIGIOUS OBEISANCE
LAYS
EARNING REDEMPTION WITH THEIR KINDNESS
HELPING CONSERVE WILDLIFE
MY HERO
Kurkure
WHAT THEY'RE ACTUALLY DOING BY FEEDING MONKEYS
WRECKING THEIR ECOLOGY, MAKING LEAF-EATERS & SEED DISPERSERS ADDICTED TO JUNK
BURP!
MAKING MONKEYS PRONE TO ROAD ACCIDENTS
PROVISIONAL FOOD HAS BEEN PROVEN TO ELEVATE STRESS LEVELS IN MONKEYS.
I WASN'T BORN A CRIMINAL. SOCIETY MADE ME ONE. NOW HAND OVER THE CAR KEYS.
LAYS

#37

SOME USES OF PALM OIL
lip gloss
Sandwich spread
Shaving cream
SOME IMPACTS OF PALM OIL
Bornean Orangutan blood
Helmeted Hornbill blood
Sumatran Rhinoceros blood

#38

BEING BLIND, WE GANGETIC DOLPHINS RELY ENTIRELY ON SONAR TO COMMUNICATE & FIND FOOD.
NOISE FROM VESSELS IN THE GANGA HAMPERS OUR ECHOLOCATION & STRESSES US OUT. THIS COULD POSE A GRAVE THREAT TO OUR ALREADY ENDANGERED SPECIES.
SO PLEASE KEEP THE GANGA NOISE-FREE.
KEEP THE GANGA NOISE-FREE!
KEEP THE GANGA NOISE-FREE!
WHRRRRRRRR
DANG. HOW CAN I MAKE MORE NOISE THAN VESSEL TRAFFIC?
TRY TWITTER, WHATSAPP OR PRIME-TIME DEBATE.

#39

NEW TITI MONKEY DISCOVERED IN COLOMBIA
The Red-Bearded Titi (Callicibus caquetensis)
TEDDY-BEAR LIKE MAMMAL DISCOVERED
The Olinguito (Bassaricyon neblina)
NEW TAPIR FOUND IN BRAZIL
The Kabomani Tapir (Tapirus kabomani)
DEFORESTATION– BRINGING SHY, ELUSIVE AMAZONIAN ANIMALS IN THE LIMELIGHT SINCE 1607!
RONAN

#40

HEY! CAN YOU PLEASE TRANQUILIZE THE CROWD FIRST?
RONAN

THREE
Mammals

WHEN THREATENED, THE PANGOLIN CURLS UP INTO A SPINY BALL OF ARMOUR.
RING
#1
WHY CAN'T THEY JUST TEXT?

Before you conclude that all my work is just gloom and doom, let me introduce you to some of the light-hearted parts of this book. I'm going reverse in evolution, starting with mammals for the simple reason that this isn't a textbook. Textbooks define mammals as vertebrates with mammary glands, hair and three middle-ear bones. What they forget to add is the characteristic quirkiness that is exclusive to our brethren! Despite being recent inhabitants of the planet in terms of evolutionary history, mammals have colonized every biome.

From the social media handles of tigers and pangolins to what it must feel like being a bat, here's a look at some of my favourite mammals from India and beyond.

#2

SOCIAL NETWORKING THE TIGER WAY

NEIGHBOURING FEMALE SENT YOU A FRIEND REQUEST.
1
SNIFF HERE TO ACCEPT.
RIVAL MALE POKED YOU.
SCRATCH HERE TO POKE BACK.
FEMALE IN HEAT POSTED ON YOUR WALL.
PISS HERE TO LIKE POST.

#3

WHY ARE YOU GRINNING LIKE A LOON?
NOT GRINNING. IT'S THE FLEHMEN RESPONSE.
THE WHAT?
SEE, A TIGRESS IN HEAT JUST PASSED BY. I'M TRYING TO INHALE HER SCENT WITH MY UPPER LIP CURLED, SO THAT HER PHEROMONES REACH MY VOMERONASAL ORGAN. THE PHEROMONES TELL THE ORGAN IF SHE IS INTERESTED OR NOT.
WHOA! SO WHAT DID HER PHEROMONES SAY TO YOUR VOMERONASAL ORGAN?
IT'S A DATE, STUDMUFFIN!
NOW YOU'RE JUST GRINNING LIKE A LOON.
YUP.

#4

pangolin
ASK ME ANYTHING, INSTAGRAM!
YOUR FAVOURITE FILM?
ANTZ
NO-BRAINER!
WHICH SUPERHERO WOULD YOU COSPLAY?
AGAIN, NO-BRAINER!
YOUR IDEA OF A GREAT KISS?
slurp!
IT'S NOT A KISS IF THERE'S NO TONGUE.
YOUR FAVOURITE MUSIC?
THE SOUND OF AN ANTHILL CRACKING OPEN TO MY CLAWS.
THE FANCIEST THING YOU'VE EVER BEEN CALLED?
"MYRMECOPHAGE!"
(Google it!)
YOUR MOST ADVENTUROUS MEAL SO FAR?
FIRE ANTS!
YOUR MESSAGE TO YOUR PREDATORS?
Boop
I'M NOT A FOOTBALL.
YOUR MESSAGE TO CRITICS & TROLLS?
GOOD LUCK GETTING THROUGH THIS BALL OF ARMOUR.
HOW DOES IT FEEL BEING THE MOST TRAFFICKED MAMMAL ON EARTH?
I FEEL LIKE ROLLING UP INTO A BALL AND CRYING.
-ROHAN

#5
ONE ADVANTAGE OF BEING ENDANGERED
HI. I'LL COME STRAIGHT TO THE POINT. WE'VE GOT TO MATE. NOW.
WHAT?! WON'T YOU EVEN TRY TO COURT ME?
LOOK, GIRL, THERE ARE ONLY TWENTY AMUR LEOPARDS ALIVE. NOW IF YOU DON'T...
OKAY, OKAY!
HA! EASIER THAN I THOUGHT!
MMMMUAH
BABY, WHERE ARE WE ON IUCN's RED LIST?

#6
USES OF THE SNOW LEOPARD'S REALLY LONG TAIL
AS A BLANKET
FOR BALANCE
SNOWBOARDING
MULTIPLE BOOKMARK
BABYSITTING
JUST SOME WARM COMPANY OVER COFFEE
GIVING ITS MATE A BOA-HUG
KINKY STUFF
PAUNCH CONCEALER

#7
IN THE FORESTS OF CENTRAL INDIA, SLOTH BEARS HAVE A PENCHANT FOR FEEDING ON MAHUA FLOWERS, WHICH BLOOM IN SPRING.
IN FACT, SLOTH BEARS ARE EVEN KNOWN TO GET INTOXICATED BY THE FERMENTED ALCOHOLIC FLAVOURS OF MAHUA!
IT'S CALLED...
THE MAHUA MACARENA!
-RAMAN

#9
WOW, GAUR!
HOW DO YOU MANAGE TO DO IT EVERY DAY?!
DO WHAT?
FIND MATCHING SOCKS?

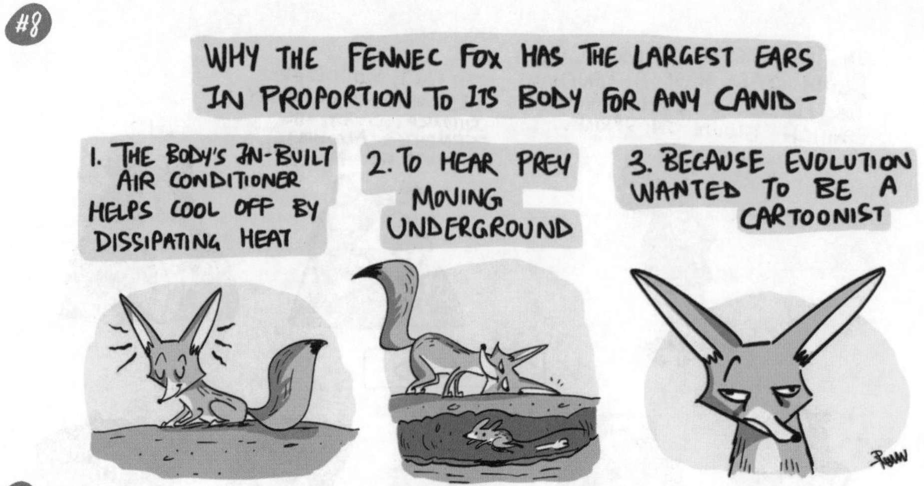
#8
WHY THE FENNEC FOX HAS THE LARGEST EARS IN PROPORTION TO ITS BODY FOR ANY CANID-
1. THE BODY'S IN-BUILT AIR CONDITIONER HELPS COOL OFF BY DISSIPATING HEAT
2. TO HEAR PREY MOVING UNDERGROUND
3. BECAUSE EVOLUTION WANTED TO BE A CARTOONIST
ROWAN

#9
CARIBOU FACTS FOR CHRISTMAS
EVEN A DAY-OLD CARIBOU CAN OUTRUN AN ATHLETE!
HOW THE HELL DID HE CLEAR THE DOPE TEST?!
THE MIGRATION OF THE PORCUPINE CARIBOU (A SUB-SPECIES OF CARIBOU) IS THE LONGEST FOR ANY LAND MAMMAL IN THE WORLD!
LADIES & GENTLEMEN, I CAN BARELY EXPRESS IN WORDS THE JOY & HONOUR OF WINNING THE 'ARCTIC TERN AWARD FOR LONGEST TERRESTRIAL MIGRATION.'
CARIBOU ARE THE ONLY DEER SPECIES IN WHICH FEMALES GROW ANTLERS!
WE CAN DO IT!
THE CARIBOU'S SUMMER COAT IS DARK BROWN AND ITS WINTER COAT IS GREYISH-WHITE.
SERIOUSLY, JULIA. THAT'S SO JULY!
CARIBOU FACE SEVERE THREATS FROM GLOBAL WARMING, CLIMATE CHANGE, HABITAT LOSS AND OIL EXPLORATION IN THE ARCTIC.
CARIBOUT CARIBOU THIS CHRISTMAS!
ROWAN

#10
THE FANTASTIC ARCTIC FOX!
THE ARCTIC FOX'S FUR HAS THE BEST INSULATION FOR ANY MAMMAL!
I REALLY APPRECIATE THE THOUGHT, MISHKA, BUT THIS IS PRACTICALLY THE MOST USELESS BIRTHDAY GIFT.
THE EGGS OF THE SNOW GOOSE ARE A FAVOURITE SNACK OF THE ARCTIC FOX, WHICH IT STEALS AND CACHES WHEN ABUNDANT.
SCRAMBLED OR FRIED?
SUNNY SIDE UP!
THE ARCTIC FOX HAS EXTREMELY KEEN HEARING AND CAN EVEN DETECT SOUNDS UNDER THE SNOW! ON DETECTING PREY, THE FOX SNOW-DIVES TO CATCH IT.
PING!
DARN FACEBOOK NOTIFICATIONS!
THE AVERAGE LITTER SIZE OF ARCTIC FOXES IS 5 TO 8, BUT IT MAY GO UP TO 25, THE LARGEST AMONG LAND CARNIVORES!
HONEY, COULD YOU FEEL THEM KICK?
YEAH. ALL 25 OF THEM.
ARCTIC FOXES OFTEN FOLLOW POLAR BEARS TO SCAVENGE ON LEFTOVERS FROM THE BEARS' MEALS.
HEY, I'VE HEARD SEALS GO REALLY WELL WITH BARBECUE SAUCE!
THE COLOUR OF THE ARCTIC FOX'S COAT CHANGES WITH SEASONS. ITS WHITE WINTER COAT BLENDS WITH THE SNOW, WHILE THE SUMMER COAT IS BROWN.
ARCTIC FOX HAUTE COUTURE
WINTER
SUMMER
SPRING
THE ARCTIC FOX'S BODY IS ROUNDED - AN ADAPTATION TO MINIMIZE HEAT LOSS. THIS MAKES IT IRRESISTIBLY CUDDLY!
SO WHAT'S FOR DESSERT?
SMACK♡
I'VE JUST HAD MINE.

#11
USES OF THE HOOLOCK GIBBON'S REALLY LONG ARMS
BRACHIATION
CANOPY GYMNASTICS
SELF-HAMMOCK
TRAINING BAR
REASSURING GROUP HUGS
REALLY EFFECTIVE 'TALK TO THE HAND' GESTURE
EASILY SCRATCHING THAT LAST ITCHY SPOT ON THE BACK
EXPRESSING AFFECTION
Daddy loves you THIIIIIIIIIIIIIS much!
USING ONE ARM AS A SELFIE-STICK, AND THE OTHER TO ENSURE EVERYONE'S IN THE FRAME.
-Rohan

#12
MACAQUE PSYCHOTHERAPY
ARE YOU STRESSED?
YES.
groom groom
THANKS, DOC!
-Rohan

#13
A TIGER'S DAILY TO-DO LIST:
① TAKE A COLD SAUNA
② GROOM
③ PATROL TERRITORY
④ RE-MARK TERRITORY
⑤ CHECK FOR LADY SCENTS
⑥ CONDUCT HERBIVORE CENSUS
⑦ MANAGE JUNGLE ECONOMY
⑧ SOMEHOW SQUEEZE IN A SIXTEEN-HOUR SNOOZE
Z
⑨ ENDORSE JUNGLE TO PAPARAZZI
⑩ ESCAPE POACHERS
⑪ SURVIVE
PHEW. FOR A MEMBER OF THE CAT FAMILY, I'M AN AWFULLY BUSY CHAP.
-RANAN

#14

THE SIX MORPHS OF THE ASIAN GOLDEN CAT

SIX ARUNACHALI GOLDEN CATS WANTED TO HAVE SOME FUN
SO THEY ALL DRANK SIX ROUNDS EACH OF APONG ONE BY ONE.
'LET'S DO SOMETHIN' CRAZY, GIRLS!' SAID ONE HIGH GOLDEN CAT
AND THE OTHER FIVE AGREED AT THE DROP OF A HAT!
OFF THEY MARCHED TO THE SALÓN AND ASKED FOR A COMPLETE MAKEOVER
MISHMI HILLS BEAUTY SALÓN
BUT THE LAST CAT IN QUEUE, IN THE NICK OF TIME, TURNED SOBER.
HIC
THE NEXT MORNING THEY MADE HEADLINES IN THE PAPERS & ON WHATSAPP-
NORTHEAST NEWS
SIX MORPHS OF THE ASIAN GOLDEN CAT DISCOVERED IN ARUNACHAL!
SIX DIFFERENT MORPHS OF THE GOLDEN CAT HAD BEEN CAMERA-TRAPPED!
CLICK
CLICK
CLICK
CINNAMON, GREY, MELANISTIC, OCELOT-SPOTTED, ROSETTED, AND THE USUAL GOLD...
THE GOLDEN ONE THAT HAD SOBERED UP SAID, 'I GUESS I'M JUST GROWING OLD.'

#15

A.M.A. WITH THE RED PANDA
@bam_boo
Ask me anything!
Type here ...
THREE THINGS YOU JUST CANNOT DO WITHOUT?
BAMBOO, BAMBOO & BAMBOO.
ISN'T IT LONELY BEING THE ONLY MEMBER IN THE FAMILY 'AILURIDAE'?
I'M A SOLITARY ANIMAL, SO IT WORKS FOR ME!
YOUR TOP BINGE-WATCH RECOMMENDATION?
WATCHING BAMBOO STEMS SWAY IN THE BREEZE. IT'S NOT JUST ENTERTAINMENT; IT'S THERAPY!
FAVOURITE CARTOON CHARACTER- KICHI, OR MASTER SHIFU?
MASTER SHIFU'S GOT THAT RED PANDA ATTITUDE; SO HIM.
FAVOURITE DRINK?
RICE BEER IN A BAMBOO BARREL. THOSE AROMAS MIXING – OH MY GOD!
CHROME OR FIREFOX?
PLEASE ASK BETTER QUESTIONS.
HOW ARE YOU RELATED TO THE GIANT PANDA?
HE'S A DISTANT RELATIVE WHO I BUMP INTO AT WEDDINGS. I'M ACTUALLY MORE CLOSELY RELATED TO WEASELS.
THEN WHY THE NAME PANDA?
IT WAS SUPPOSED TO BE 'POONYA' ('BAMBOO EATER' IN A NEPALI DIALECT). BUT, EUROPEANS, I TELL YOU!
FAVOURITE FILM?
GUNJAN MENON'S 'THE FIREFOX GUARDIAN'. P.S.- I'M IN IT!
YOUR IDEA OF A PERFECT DATE?
GNAWING ON THE SAME BAMBOO SHOOT FROM OPPOSITE ENDS, ENDING IN A KISS.
YOUR MOST TRUSTED COMPANION?
MY BUSHY TAIL.
WHAT DOES THE COLOUR RED MEAN TO YOU?
En
AS OF NOW, MY IUCN STATUS: ENDANGERED!
-Rohan

#16

4 ADVANTAGES OF BEING A BAT

IN-BUILT SUPERHERO CAPE THAT ACTUALLY WORKS.
IT'S ALWAYS TWO THUMBS UP TO EVERYTHING.
MORE?
SWEAR AT PEOPLE IN ULTRASONIC, AND GET AWAY WITH IT.
@#* *%$!
MAKING OUT IN PUBLIC IS NEVER A HASSLE.
mmm
mmm
smooch
kiss
slurrrp
-Rohan

LADIES & GENTLEMEN, MEET THE PAINTED BAT, INDIA'S MOST STRIKINGLY COLOURED BAT!
BUT THE QUESTION IS— WHY ARE YOU COLOURED ORANGE?
CAN YOU GUESS?
DO YOU EAT A LOT OF ORANGES?
NOPE! I USE ECHO LOCATION TO HUNT WINGED INSECTS.
ARE YOU A SUPPORTER OF THE SAFFRON PARTY?
UGH! WITH THEIR PATHETIC ENVIRONMENTAL TRACK RECORD? NEVER!
ARE YOU A FAN OF THE NETHERLANDS SOCCER TEAM?
I SUPPORT MY HOME TEAM, INDIA!
CHHETRI 11
ARE YOU A FAN OF TIGERS?
WE HAVE A MUTUAL ADMIRATION CLUB, BUT THAT'S NOT WHY I'M ORANGE...
MY ORANGE COLOUR IS AN ADAPTATION FOR CAMOUFLAGE, WHEN ROOSTING IN DRIED BANANA LEAVES.
WHOA, NEAT!
-Rohan

#17

LESSONS FROM THE INDIAN JACKAL DAD

NEVER BEG OR BORROW. ALWAYS STEAL.

HEY!

IF YOU BELIEVE YOU CAN CATCH JUST ABOUT ANYTHING, YOU CAN CATCH JUST ABOUT ANYTHING.

NEVER ABANDON A CHASE ONCE YOU'VE STARTED IT.

NEVER BE ASHAMED OF SWITCHING TO BERRIES IN TOUGH TIMES.

NEVER EVER CHEAT ON YOUR MATE.

IT'S PERFECTLY OKAY TO THROW UP IN FRONT OF THE KIDS.

ACK!

#18

USES OF THE FLYING SQUIRREL'S PATAGIA (GLIDING MEMBRANES
ESCAPING PREDATORS
AVOIDING SMALL TALK AT PARTIES
SO HOW'S THE-
MAKING A QUICK ESCAPE FROM A POTENTIALLY DISASTROUS DATE
WHY, OF COURSE I READ CHETAN BH-
HEY!
ENACTING YOUR OWN SUPERHERO JINGLE
NANANA NA NANANANA FLYING SQUIRREL!
PARAGLIDING WITHOUT HAVING TO GO TO BIR BILLING
WHEEE!
REALLY, REALLY FLUFFY HUGS

#19

STUDIES SUGGEST THAT HYBRIDIZATION WITH DOMESTIC CATTLE IS THE BIGGEST THREAT TO THE POPULATION OF THE ENDANGERED ASIATIC WILD BUFFALO!
YOUR MATCHES
Indu
Suchitra
Gowri
YOU MEAN I HAVE TO UNMATCH WITH ALL THESE COWS NOW?

#20

SLEEP (A poem by Sloth)
THE CHEETAH HAS EVOLVED INTO AN ATHLETE, THE ORANGUTAN HAS LEARNED TOOL-USE, MAN HAS EVOLVED INTO A MASTERMIND, BUT I'VE JUST EVOLVED TO SNOOZE.
Z.
I HANG FROM BRANCHES UPSIDE DOWN, WHILE MY LONG CLAWS GRASP THEM TIGHT, A BUNCH OF LEAVES IS ALL I NEED FOR MY MODEST APPETITE.
Z.
EIGHT HOURS A NIGHT A MAN SLEEPS FOR, A MERE THIRD OF HIS LIFE. BUT I SPEND JUST ABOUT MY WHOLE DAY IN BED, AS DO MY KID & MY WIFE.
Z.
TO YOU IT MAY SEEM LIKE A TRIVIAL TASK, BUT TO ME IT IS PROFOUNDLY DEEP— YOU SEE, REST ISN'T JUST A REQUISITE, THE TRUE MEANING OF LIFE IS SLEEP.
Z.

#21

SOME JOYS OF BEING A KANGAROO
HOP AWAY FROM LIFE'S PROBLEMS AT WILL.
STUFF YOUR BABY IN YOUR POUCH IF IT WHINES IN PUBLIC.
PENTAPEDAL LOCOMOTION: WHEN WALKING ON FOURS GETS TOO MAINSTREAM.
REVERSED FASHION GENDER BIAS: MEN MUST DEAL WITH HAVING NO POCKETS.
CAN I KEEP MY PHONE IN THERE?
WHAT'S THE MAGIC WORD, DEAR?

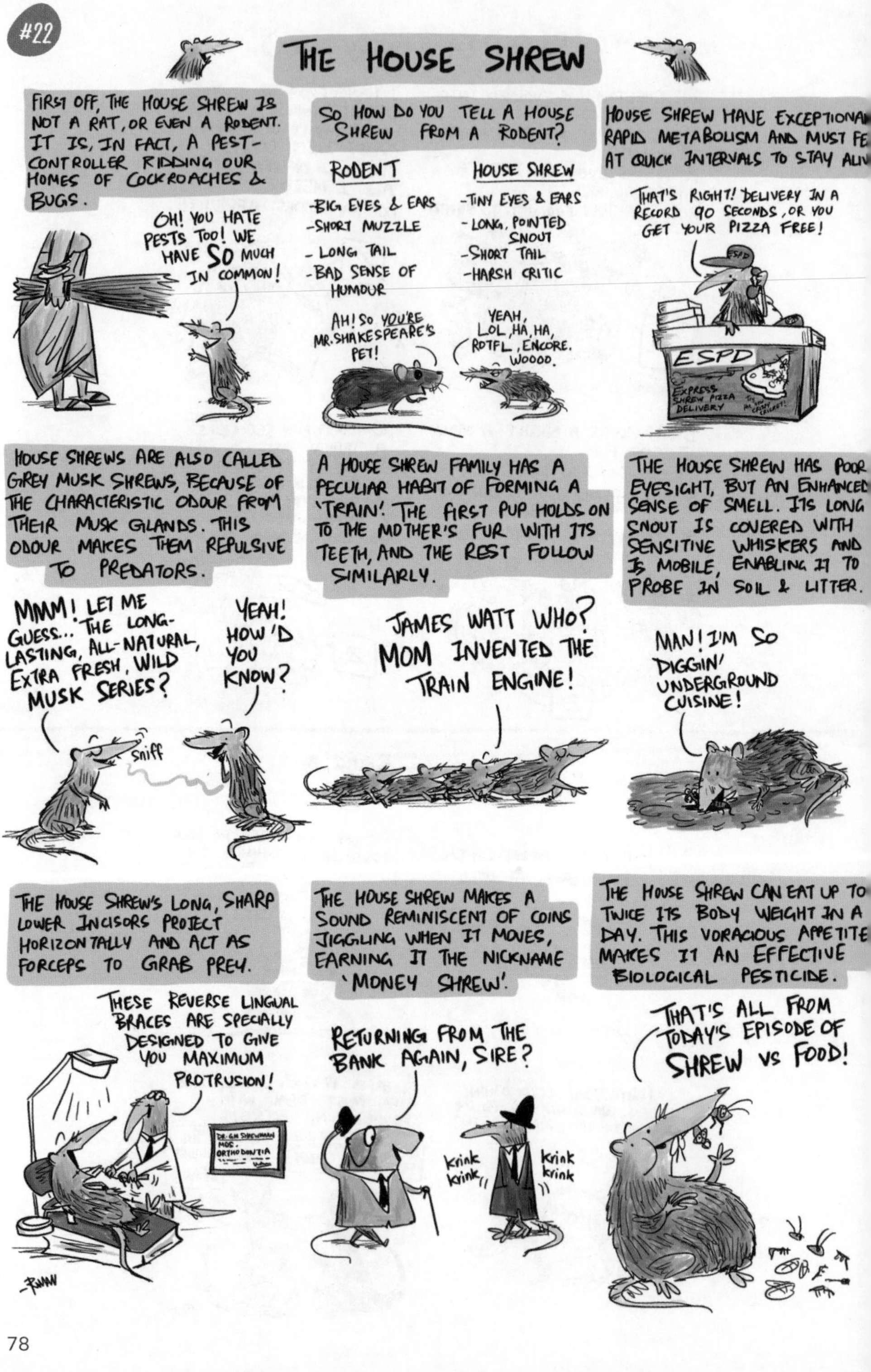
#22
THE HOUSE SHREW
FIRST OFF, THE HOUSE SHREW IS NOT A RAT, OR EVEN A RODENT. IT IS, IN FACT, A PEST-CONTROLLER RIDDING OUR HOMES OF COCKROACHES & BUGS.
OH! YOU HATE PESTS TOO! WE HAVE SO MUCH IN COMMON!
SO HOW DO YOU TELL A HOUSE SHREW FROM A RODENT?
RODENT
-BIG EYES & EARS
-SHORT MUZZLE
- LONG TAIL
-BAD SENSE OF HUMOUR
HOUSE SHREW
-TINY EYES & EARS
- LONG, POINTED SNOUT
-SHORT TAIL
-HARSH CRITIC
AH! SO YOU'RE MR. SHAKESPEARE'S PET!
YEAH, LOL, HA, HA, ROTFL, ENCORE. WOOOO.
HOUSE SHREW HAVE EXCEPTIONA
RAPID METABOLISM AND MUST FE
AT QUICK INTERVALS TO STAY ALIV
THAT'S RIGHT! DELIVERY IN A RECORD 90 SECONDS, OR YOU GET YOUR PIZZA FREE!
ESPD
EXPRESS SHREW PIZZA DELIVERY
HOUSE SHREWS ARE ALSO CALLED GREY MUSK SHREWS, BECAUSE OF THE CHARACTERISTIC ODOUR FROM THEIR MUSK GLANDS. THIS ODOUR MAKES THEM REPULSIVE TO PREDATORS.
MMM! LET ME GUESS... THE LONG-LASTING, ALL-NATURAL, EXTRA FRESH, WILD MUSK SERIES?
sniff
YEAH! HOW'D YOU KNOW?
A HOUSE SHREW FAMILY HAS A PECULIAR HABIT OF FORMING A 'TRAIN'. THE FIRST PUP HOLDS ON TO THE MOTHER'S FUR WITH ITS TEETH, AND THE REST FOLLOW SIMILARLY.
JAMES WATT WHO? MOM INVENTED THE TRAIN ENGINE!
THE HOUSE SHREW HAS POOR EYESIGHT, BUT AN ENHANCED SENSE OF SMELL. ITS LONG SNOUT IS COVERED WITH SENSITIVE WHISKERS AND IS MOBILE, ENABLING IT TO PROBE IN SOIL & LITTER.
MAN! I'M SO DIGGIN' UNDERGROUND CUISINE!
THE HOUSE SHREW'S LONG, SHARP LOWER INCISORS PROJECT HORIZONTALLY AND ACT AS FORCEPS TO GRAB PREY.
THESE REVERSE LINGUAL BRACES ARE SPECIALLY DESIGNED TO GIVE YOU MAXIMUM PROTRUSION!
DR. G.M. SHREWMAN MDS. ORTHODONTIA
THE HOUSE SHREW MAKES A SOUND REMINISCENT OF COINS JIGGLING WHEN IT MOVES, EARNING IT THE NICKNAME 'MONEY SHREW'.
RETURNING FROM THE BANK AGAIN, SIRE?
krink krink
krink krink
THE HOUSE SHREW CAN EAT UP TO TWICE ITS BODY WEIGHT IN A DAY. THIS VORACIOUS APPETITE MAKES IT AN EFFECTIVE BIOLOGICAL PESTICIDE.
THAT'S ALL FROM TODAY'S EPISODE OF SHREW VS FOOD!

#23

SOME FACTS ABOUT DHOLES
DHOLES, OR ASIATIC WILD DOGS, ARE HIGHLY SOCIAL ANIMALS, LIVING IN CLANS OF AS MANY AS 40 DOGS.
ACCEPT ALL, BABY!
YOU HAVE 40 NEW FRIEND REQUESTS
Dan ACCEPT
Rita ACCEPT
Pupsy ACCEPT
DHOLES DIFFER FROM OTHER CANIDS IN HAVING FEWER MOLARS AND MORE TEATS.
HOW MUCH FOR THIS ONE?
DHOLES VOCALIZE BY MAKING UNIQUE 'COO-OOO' WHISTLES, UNLIKE ANY OTHER CANID.
WAIT. WHY AM I ALWAYS THE REFEREE?!
'CAUSE NOBODY ELSE CAN WHISTLE LIKE YOU DO!
DHOLE CLANS HAVE LESS DOMINANCE AND MORE EQUALITY & TOLERANCE FOR OTHER MEMBERS, COMPARED TO WOLF PACKS.
OH, AND NO 'SIR', PLEASE! CALL ME DOUG.
SURE, SIR... OOPS! DOUG... tee hee!
CEO
DHOLE CLANS HAVE MULTIPLE BREEDING FEMALES.
THEY'RE ALL ON MATERNITY LEAVE.
TINA
SARA
DHOLES BELIEVE IN CO-OPERATION. EACH CLAN MEMBER FEEDS THE NURSING MOTHERS & PUPS.
SO WHOOOO WANTS DEER RUMP AND WALNUT FUDGE?
AWW, UNCLE, YOU DIDN'T HAVE TO!
DHOLES HUNT IN WELL-ORGANIZED HUNTING PARTIES. THIS MAKES THEM CAPABLE OF TAKING DOWN PREY MUCH LARGER THAN THEMSELVES.
TAG TEAM CHAMPIONS
HUNTING IN LARGE GROUPS ALSO GIVES THEM A BETTER SUCCESS RATE THAN OTHER CARNIVORES.
GREETINGS, FRIENDS! I AM DHOLE, YOUR FACILITATOR, AND I WELCOME YOU ALL TO MY TEAM-BUILDING WORKSHOP!
SUCCESS & TEAM WORK
DHOLES ARE RUTHLESS HUNTERS AND START EATING THEIR PREY EVEN BEFORE IT'S DEAD!
AAOW! LEARN SOME TABLE ETIQUETTE!
BUT DHOLES ARE DARLINGS TOO. UNLIKE MOST OTHER CANIDS, THEY ALWAYS LET THEIR PUPS EAT FIRST!
THEN 'SNAP!' LITTLE RON BIT THE SAMBAR'S LEG... AND 'THUD!' THE GIANT TUMBLED... AND THEN 'MUNCH MUNCH MUNCH' WENT LITTLE RON!

#24
ELEPHANT DUNG
A CRAPLOAD OF MAGIC!
ELEPHANTS ARE GLUTTONS AND POOR DIGESTERS. MOST SEEDS THEY CONSUME ARE PASSED OUT INTACT.
PHEW! I COULD ALMOST FEEL A WHOLE TREE GROW INSIDE ME!
ELEPHANTS ARE CAPABLE OF DISPERSING SEEDS ACROSS VERY LARGE DISTANCES.
OH, HERE'S MY PLANE. SEE YOU IN HAWAII!
AS ELEPHANTS TRANSPORT SEE OF SEVERAL PLANT SPECIES, THE DUNG CAN REJUVENATE AN ENTIRE HABITAT!
MY HOBBIES? WELL... I'M A BIT OF A HORTICULTURE AFICIONADO.
ELEPHANT DUNG PROVIDES AN EXCELLENT ENVIRONMENT FOR GERMINATION.
WHAT WOULD YOU LIKE TO BE REBORN AS?
A SEED IN A PILE OF JUMBO POO.
ELEPHANT DUNG IS A RICH SOURCE OF MINERALS FOR ANIMALS AND A GREAT NUTRIENT CYCLER FOR THE SOIL.
WELCOME TO THE WORLD'S FIRST OPENCAST MINE WITH ZERO BIO-HAZARD!
SEVERAL ANIMALS FEED ON ELEPHAN DUNG, INCLUDING BIRDS, MONKEYS AND A WHOLE BUNCH OF INSEC SUCH AS DUNG BEETLES.
MAY I RECOMMEND THE ASSORTED SALAD BY OUR HEAD CHEF?
ELEPHANT DUNG IS HOME NOT ONLY TO INVERTEBRATES BUT ALSO TO A FEW SPECIES OF FROGS!
I THINK IT'S GREAT, HONEY! WHAT MAKES YOU CALL THIS PLACE A DUMP?
THUS, A PILE OF ELEPHANT DUNG IS AN ECOSYSTEM IN ITSELF!
WE HEREBY APPOINT YOU AS RANGER OF DUNGBULLA NATIONAL PARK.
REMEMBER, SON. ALWAYS GIVE BACK TO NATURE MORE THAN YOU TAKE FROM HER.
ROHAN

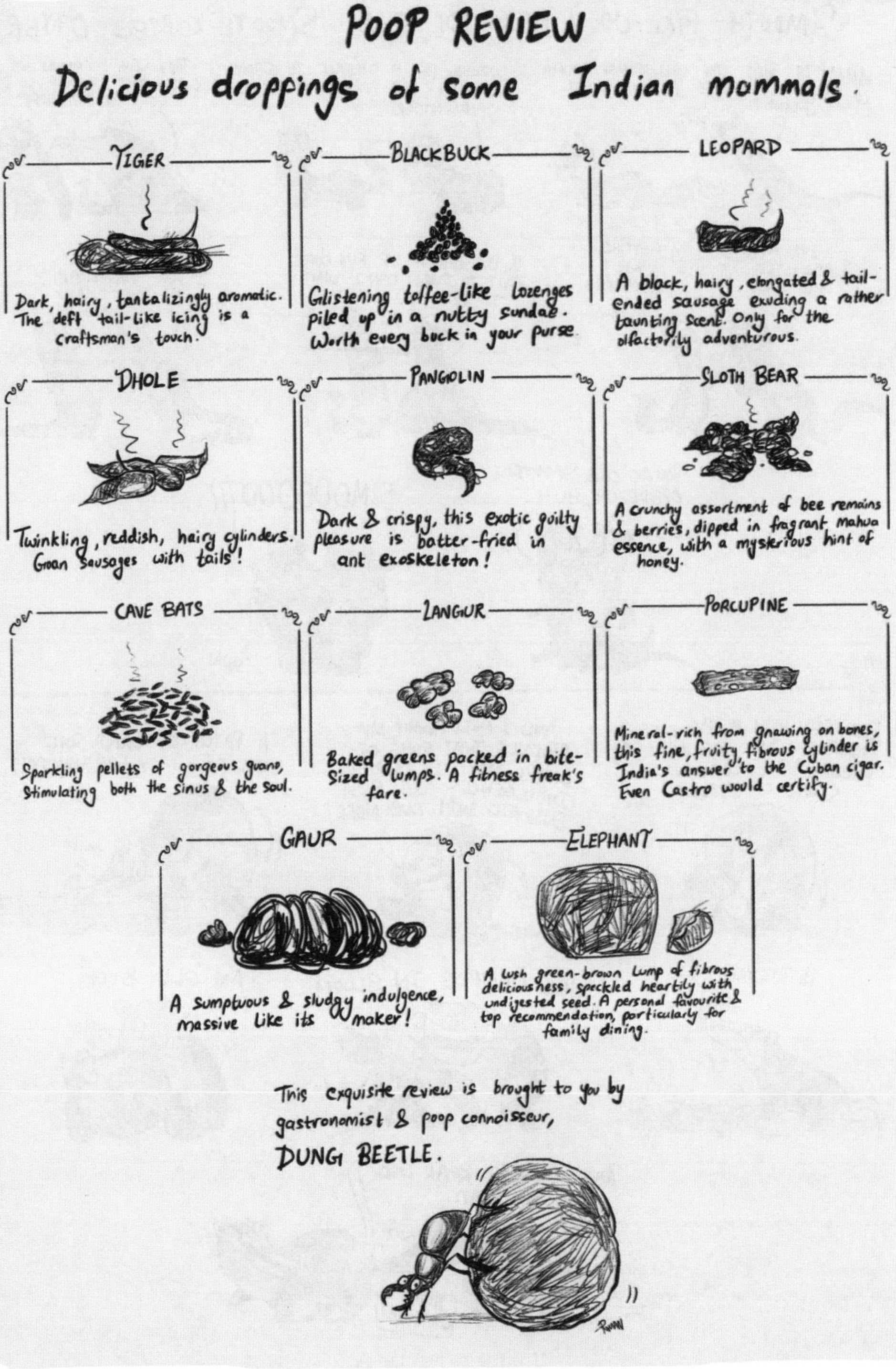
POOP REVIEW
Delicious droppings of some Indian mammals.
TIGER
Dark, hairy, tantalizingly aromatic. The deft tail-like icing is a craftsman's touch.
BLACKBUCK
Glistening toffee-like lozenges piled up in a nutty sundae. Worth every buck in your purse.
LEOPARD
A black, hairy, elongated & tail-ended sausage exuding a rather taunting scent. Only for the olfactorily adventurous.
DHOLE
Twinkling, reddish, hairy cylinders. Goan sausages with tails!
PANGOLIN
Dark & crispy, this exotic guilty pleasure is batter-fried in ant exoskeleton!
SLOTH BEAR
A crunchy assortment of bee remains & berries, dipped in fragrant mahua essence, with a mysterious hint of honey.
CAVE BATS
Sparkling pellets of gorgeous guano, stimulating both the sinus & the soul.
LANGUR
Baked greens packed in bite-sized lumps. A fitness freak's fare.
PORCUPINE
Mineral-rich from gnawing on bones, this fine, fruity, fibrous cylinder is India's answer to the Cuban cigar. Even Castro would certify.
GAUR
A sumptuous & sludgy indulgence, massive like its maker!
ELEPHANT
A lush green-brown lump of fibrous deliciousness, speckled heartily with undigested seed. A personal favourite & top recommendation, particularly for family dining.
This exquisite review is brought to you by gastronomist & poop connoisseur,
DUNG BEETLE.

#26

SMOOTH PICK-UP LINES BY THE SMOOTH-COATED OTTER

MMM! IS THAT THE ALLURING AROMA OF FISH FILLET OR IS IT YOUR SCENT GLAND?
sniff
LOOKING FOR A SURFACE TO SPRAY SCENT ON? MY FACE WANTED TO VOLUNTEER.
THEY SAY DIAMONDS ARE A GIRL'S BEST FRIENDS, SO I GOT YOU MY NOSE.
I'M AFRAID I'M GETTING ADDICTED TO THE WEB! THE ONE IN YOUR MUSHY PAWS.
I DISAPPROVE OF FUR COATS, BUT I SURE CAN'T WAIT TO TRY YOU ON.
IUCN IS RIGHT. I AM 'VULNERABLE'... TO YOUR CHARM.
YOU'RE ONE SMOOTH OPERATOR, HUH!
SMOOOOOOOTH.

#27

EQUIPPED WITH A PIG-LIKE SNOUT, I, THE HOG BADGER, POSSESS AN EXCEPTIONAL SENSE OF SMELL!
AND I MUST ADMIT, MY FRIENDS, THAT SOME OF LIFE'S MOST PLEASURABLE INDULGENCES ARE BEST ENJOYED WITH YOUR NOSE!
A PATCH OF DAMP SOIL FULL OF JUICY EARTHWORMS...
sniff
PETRICHOR...
sniff
WILD ORCHIDS IN BLOOM...
AN OLD BOOK...
sniff
SH PRATER
INDIAN ANIMALS
YOUR MATE'S REAR END IN HEAT...
sniff
AHH, CAMELIA!

#28
MYTH VERSUS FACT:
WOMBATS AND THE AUSTRALIAN BUSHFIRES
MYTH: WOMBATS MAKE SHELTERS FOR ANIMALS AFFECTED BY BUSHFIRES.
FACT: WOMBATS DIG NETWORKS OF BURROWS WHICH MAY BE USED BY OTHER ANIMALS AS SHELTER.
MYTH: WOMBATS DIG WELLS TO OFFER WATER TO BUSHFIRE VICTIMS.
FACT: WOMBATS DIG DEEPER BURROWS DURING DROUGHTS, EXPOSING GROUNDWATER. THESE 'SOAKS' MAY BE USED BY OTHER ANIMALS FOR A DRINK.
MYTH: WOMBATS AREN'T AS HEROIC AS BEING PORTRAYED TO BE.
FACT: A WOMBAT WOULD MAKE A BETTER PRIME MINISTER THAN SCOTT MORRISON.
WOMBAT FOR PM

FOUR

Birds

#1

HOOPOE DRESSING UP ON A MONDAY MORNING
HOOPOE DRESSING UP ON A FRIDAY NIGHT

Birds have been the gateway to the world of wildlife for many like me, simply because they happen to be everywhere! From exemplary parenting to burning the dance floor, from record-breaking dives to rave fashion, birds can do it all. They exude mischief and character like no other class of animals. No wonder this section is the longest in the book!

#2

LOOK! UP IN THE SKY!!
IS IT A PLANE?!
IS IT SUPERMAN?!
NO, YOU FOOLS! IT'S A BIRD!
DIVING AT A WHOPPING 390 KMPH, THE FASTEST CREATURE ON EARTH...
THE PEREGRINE FALCON!™
BIG FAN OF YOUR WORK, SIR!

#3

WHAT DOES IT TAKE TO BE A GREAT DRUMMER?
-By the Flameback Woodpecker
PERSEVERANCE
AN INSTINCTIVE SENSE OF RHYTHM
TAP TAP TAPPITY TAP!
TONS OF TALENT
THE FLEXIBILITY TO ADAPT TO ANY INSTRUMENT
HOURS OF PRACTICE
BUT MOST IMPORTANTLY... A REALLY SEXY BANDANA

#4

MEET THE BAR-HEADED GEESE, THE HIGHEST FLYING BIRDS IN THE WORLD, FLYING OVER THE MIGHTY HIMALAYAS ON THEIR WAY TO INDIA...
ALRIGHT, FOLKS! LET'S SHOW EVEREST WHO'S BOSS!
WHEEEEE!
BAR-HEADED GEESE. RAISING THE BAR EACH YEAR!

#5

CUCKOO CHARTBUSTERS FOR YOUR MONSOON PLAYLIST

KOOHOO! KOOHOO!
THE ASIAN KOEL'S LOUD & ENERGIZING POP HIT. PLEASE DO NOT USE EARPHONES FOR THIS ONE!
pipee<pipee<pipee<
PEE-PEEUWI! PEE-PEEUWI!
THE COMMON HAWK-CUCKOO'S 'RAAG MALHAAR' COVER. INDIANS BELIEVE THAT IT SUMMONS THE RAIN, WHILE GRUMPY BRITISHERS CALL IT THE 'BRAINFEVER'!
WOO>OOH<OOH>OOP
THE INDIAN CUCKOO'S SOULFUL, REVERBERATING MELODY. AS MUCH A TREAT TO THE HINDUSTANI CLASSICAL PATRON'S EAR AS TO A D.J.'s!
PEE-EW-piu!
THE PIED CUCKOO'S FLUTY OUTBURSTS. A FINE FUSION OF INDIAN INSTRUMENTS WITH VOCALS FROM THE VISITOR'S NATIVE LANDS IN AFRICA.
pi<pi<pi<pi<pip!
THE DRONGO-CUCKOO'S RISING WHISTLE, APTLY PLACED IN THE BEST-SELLING ALBUM 'ASCENSION'.
PEE-pi pi-PEE<
PEE-pi pi-PEE<
PEE-pi pi-PEE!
THE GREY-BELLIED CUCKOO'S YOUTHFUL YODEL. THE PERFECT SONG FOR THOSE CRAVING FOR SOME DEV ANAND X KISHORE DA NOSTALGIA!

#6

MAKING A V-DAY GREETING CARD: A TUTORIAL BY THE BARN OWL
① CUT OUT A HEART SHAPE ON WASTE PAPER.
② WRITE YOUR MESSAGE.
③ INSERT YOUR HEART-SHAPED FACE & PRESENT IT TO HER.
④ SCORE!
OWL BE YOUR VALENTINE
OWL BE YOUR VALENTINE

#7
SOME FACTS ABOUT HORNBILLS
THE FAMILY NAME OF HORNBILLS, 'BUCEROTIDAE', MEANS 'COW HORN' IN GREEK.
BE-YOUR-FAVOURITE-BIRD DAY
I'M A HORNBILL!
HORNBILLS ARE THE ONLY BIRDS WITH FUSED ATLAS & AXIS VERTEBRAE: AN ADAPTATION FOR CARRYING THE HEAVY BILL.
PHYSIOTHERAPY FOR HORNBILLS
NECK STRAIGHT... LIFT...LIFT... GOOD!
HORNBILLS ARE MONOGAMOUS, AND MOST NEST IN TREE CAVITIES WHERE THE FEMALE SEALS HERSELF & HER EGGS USING A MIXTURE OF MUD & FECES, FOR ADDED PROTECTION FROM PREDATORS.
LOOK, WE'VE HAD THIS TALK. I NEED MY OWN SPACE.
DURING NESTING, THE FEMALE AND THE CHICKS ARE ENTIRELY DEPENDENT ON THE MALE HORNBILL FOR SURVIVAL.
WHAT? YOU FORGOT MY LIP GLOSS AGAIN?!
THE CASQUE OF THE HORNBILLS (A HOLLOW STRUCTURE LOCATED ON THE UPPER BEAK) FUNCTIONS AS A RESONATOR FOR THEIR CALLS.
VOTE FOR ME!
CLAP CLAP CLAP
THE CASQUE OF THE HELMETED HORNBILL IS FILLED WITH IVORY, AND IS USED LIKE A BATTERING RAM IN FIGHTS OVER MATES AND TERRITORY.
HORNBILL TEKKEN 3D
K.O.
BINOCULAR VISION ALLOWS HORNBILLS TO SEE THEIR OWN BILL TIPS, ENABLING HANDLING OF FOOD WITH PRECISION.
SIX MINUTES & 31...32... 33 SECONDS! YOU'VE BEATEN YOUR PREVIOUS BEST, DUDE!
AS HORNBILLS HAVE SHORT TONGUES, THEY MUST TOSS FOOD AT THE TIP OF THEIR BEAKS IN ORDER TO SWALLOW IT!
MY GOODNESS! WHAT TABLE MANNERS IS THAT KID LEARNING?
TOSS
TOSS
HORNBILLS ARE INDISPENSABLE SEED DISPERSERS OF FIGS AND A VARIETY OF FRUIT TREES, CAPABLE OF REGENERATING ENTIRE FORESTS!
IVORY ISN'T WHITE GOLD. THIS IS!
SPLAT
SOME AFRICAN HORNBILLS SHARE A SYMBIOSIS WITH DWARF MONGOOSES. THE HORNBILL EATS INSECTS DISTURBED BY THE FORAGING MONGOOSES, AND IN TURN WARNS THEM OF APPROACHING PREDATORS.
THANKS FOR DINNER, GUYS! BOOZE IS ON ME TONIGHT.
YAY!!
YAY!!
GLUG

#8

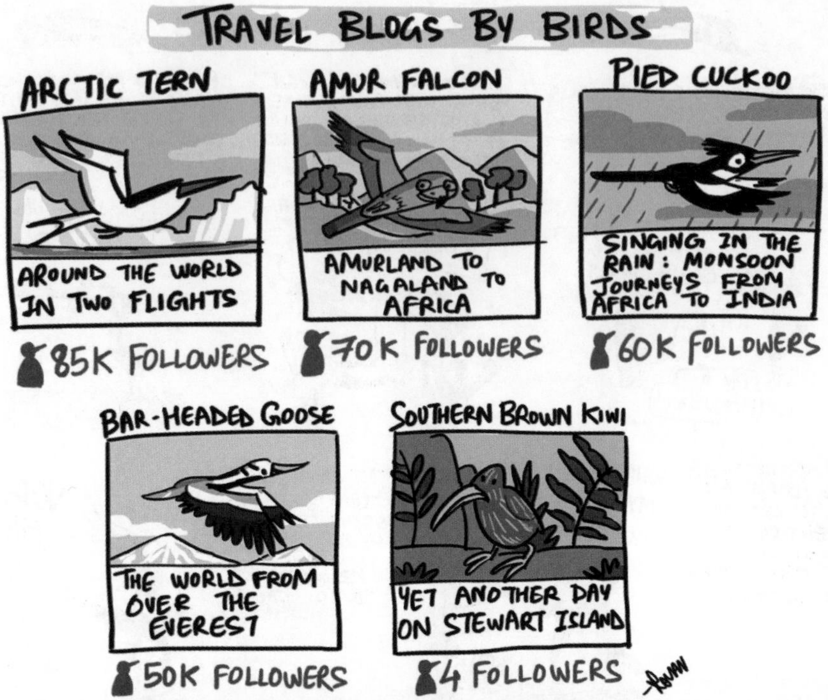
TRAVEL BLOGS BY BIRDS
ARCTIC TERN
AROUND THE WORLD IN TWO FLIGHTS
85K FOLLOWERS
AMUR FALCON
AMURLAND TO NAGALAND TO AFRICA
70K FOLLOWERS
PIED CUCKOO
SINGING IN THE RAIN: MONSOON JOURNEYS FROM AFRICA TO INDIA
60K FOLLOWERS
BAR-HEADED GOOSE
THE WORLD FROM OVER THE EVEREST
50K FOLLOWERS
SOUTHERN BROWN KIWI
YET ANOTHER DAY ON STEWART ISLAND
4 FOLLOWERS
-Rohan

#9

ARCTIC TERNS TELLING STORIES FROM THE NORTH
GROWL!!
GRRR...
ARCTIC TERNS TELLING STORIES FROM THE SOUTH
PLOP
PLOP
PLOP
Rohan

#10

THE MANY ORIGINAL COMPOSITIONS OF THE ORANGE-HEADED THRUSH

♪WEE-TOO-TIWOOO! ♫
♪TICH-CHEE-WUEE-WEE! ♫
♪TRR-CHU-CHU-CHUEE-WEE! ♫
♪TEE-WOO-WUEEWOO-WEE! ♫
UNLIKE BOLLYWOOD'S INEPT MUSICIANS, I DESPISE REMAKES.
-Rohan

#11

Dear Dad!
Remember me? The largest egg in your clutch last monsoon?
That hatched first, and then somehow, all the other eggs disappeared?
The little brat who ate so much that you & Mum could never take a break?
And who grew even larger than you?
Well, Dad, it's time I told you - I'm not a Jungle Babbler. I was dropped into your nest by my real parents, Pied Cuckoos.
The other eggs didn't disappear, Dad. I kicked them out while you were away.
But had it not been for you, I'd never have grown into a beautiful Pied Cuckoo, and found a charming mate!
Dad, you'll be really pleased to know that your little brat put her own egg in a Jungle Babbler's nest today!
PLOP!
Thanks for everything, Dad! Happy (Foster) Fathers' Day!
-Rohan

#12

THE WINGS OF A CORMORANT ARE DESIGNED TO DECREASE BUOYANCY & ACCELERATE THE PURSUIT OF AQUATIC PREY.
THIS COMES AT A SMALL PRICE. ITS WINGS, UNLIKE THOSE OF MOST OTHER WATERBIRDS, ARE NOT WATER-REPELLANT.
AS A RESULT, CORMORANTS CAN BE SEEN SPENDING LONG HOURS HOLDING THEIR WINGS OPEN IN THE SUN, TO DRY THEIR FEATHERS AFTER A DIVE.
WIDER, SHAHRUKH, WIDER!

#13

HEY! I HAVEN'T CHANGED MY PROFILE PICTURE IN MONTHS!
f
UPLOAD NEW PROFILE PICTURE?
YES
f
DRAG TO REPOSITION
f
DRAG TO REPOSITION
f
DRAG TO REPOSITION
WHAT THE CASQUE!
I GUESS I'LL JUST UPLOAD A FRONT PROFILE.
GREAT HORNBILL UPLOADED A NEW PROFILE PICTURE
Who is that?
WHAT is that?
Is everything okay?

#14

WHY SOME TROPICAL BIRDS HANG AROUND IN MIXED-SPECIES FLOCKS
THERE'S ALWAYS A SENTRY LOOKING OUT FOR PREDATORS
!
EASE IN LOCATING FOOD SOURCES
LIVE JAMMING EVERY DAY!
TAP-TAP-TATAP-TAP!
IT'S ALWAYS GREAT TO EXPERIENCE NEW CULTURES
CACKLE
CACKLE
CACKLE
CACKLE!
UMM.. CACKLE!
BOOZE COSTS ARE SPLIT EQUALLY
TO US!

#15

POND HERON IN NON-BREEDING PLUMAGE
BOUGHT THIS AT THE DECATHLON CLEARANCE SALE JUST FOR ₹ 99!!
POND HERON IN BREEDING PLUMAGE
IT'S A SABYASACHI ORIGINAL.
BRAHMINY STARLING
BRAHMINY STARLING BEFORE THE MORNING CUP OF COFFEE

#16
LITTLE THINGS YOU AND I CAN DO TO SAVE SPARROWS
① PUT UP A NEST BOX. MAKE SURE IT'S WATERPROOF.
② ADVERTISE!
NO INTRUSION BY LANDLORD!
FREE Wifi
NO RENT FOR COUPLES!
③ PLACE A BIRD BATH. SPARROWS LOVE POOLSIDE VILLAS.
④ DISCOURAGE CATS FROM APPROACHING.
⑤ SUBTLY NUDGE THE COUPLE TO MAKE BABIES.

#17
I'VE NEVER CARED ABOUT DRESSING UP.
THE FIRST TIME I PROFESSED MY LOVE TO A GIRL, I WAS DRESSED IN A MUD-STAINED SCHOOL UNIFORM.
I WALKED INTO MY EDITOR'S OFFICE IN MY SHORTS TO HAVE MY CARTOONS PUBLISHED.
WHEN I PROPOSED MARRIAGE, I WORE BATHROOM SLIPPERS TO THE DATE.
BUT TODAY, I'M GOING TO DRESS UP.
I HAVE A VERY SPECIAL DATE AND SHE'S COME A VERY LONG WAY. I'D BETTER PUT ON MY BEST SHOW!
ONE MOMENT, LOVE!
Chew-ip!
THERE SHE IS!
ALL THE WAY FROM THE MOUNTAINS OF CENTRAL ASIA TO MY BALCONY LIKE EVERY YEAR...
Chew-ip!
THE HUME'S WARBLER!
MY, MY! LOOK WHO HASN'T AGED ONE BIT!
Chew-ip!

#18

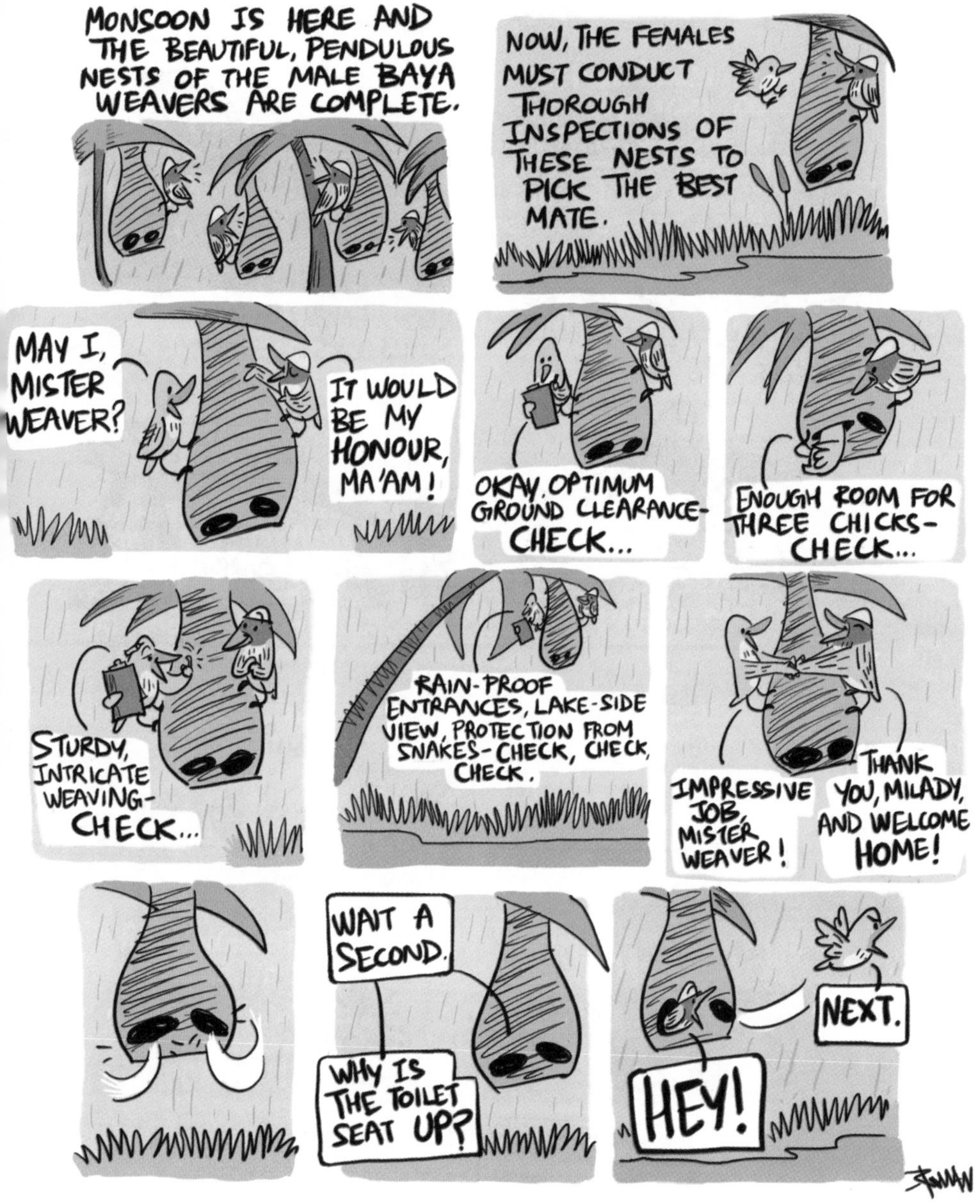
MONSOON IS HERE AND THE BEAUTIFUL, PENDULOUS NESTS OF THE MALE BAYA WEAVERS ARE COMPLETE.
NOW, THE FEMALES MUST CONDUCT THOROUGH INSPECTIONS OF THESE NESTS TO PICK THE BEST MATE.
MAY I, MISTER WEAVER?
IT WOULD BE MY HONOUR, MA'AM!
OKAY, OPTIMUM GROUND CLEARANCE- CHECK...
ENOUGH ROOM FOR THREE CHICKS- CHECK...
STURDY, INTRICATE WEAVING- CHECK...
RAIN-PROOF ENTRANCES, LAKE-SIDE VIEW, PROTECTION FROM SNAKES- CHECK, CHECK, CHECK.
IMPRESSIVE JOB, MISTER WEAVER!
THANK YOU, MILADY, AND WELCOME HOME!
WAIT A SECOND.
WHY IS THE TOILET SEAT UP?
HEY!
NEXT.

#19
THE RUDDY TURNSTONE GETS ITS NAME FROM ITS PECULIAR FEEDING HABIT-
TURNING STONES OVER TO CATCH PREY HIDING UNDERNEATH.
JUICY TITBITS LIKE CRABS, OTHER CRUSTACEANS AND INSECTS ARE CAUGHT THIS WAY...
BUT IT CAN OFTEN LEAD TO VERY EMBARRASSING SITUATIONS.
UHM... SORRY!

#20
DIP!
WHACK!
WHACK!
GULP!
FOR MORE FISHING TUTORIALS, SUBSCRIBE TO MY YOUTUBE CHANNEL NOW!

#21

The Black Drongo

MESS WITH A GRIZZLY, MESS WITH A WEASEL, MESS WITH THE BUFFALOES OF OKAVANGO,
MESS WITH ANY DARN CREATURE ON EARTH, BUT NEVER MESS WITH THE BLACK DRONGO.
HE DOESN'T CARE WHAT SIZE YOU ARE, OR WHETHER YOU HAVE FEATHERS OR FUR,
IF YOU VENTURE TOO CLOSE TO THE DRONGO'S NEST, THERE'LL BE TROUBLE FOR YOU, GOOD SIR!
@#%!
RELENTLESSLY HE'LL DIVE-BOMB YOU, UNTIL YOU'RE MILES AWAY FROM HIS BROOD,
WHACK!
SNAP!
AGGRESSIVE, FORMIDABLE, FEARLESS, GALLANT- BEHOLD THIS FORK-TAILED DUDE!
IT DOESN'T MATTER IF YOUR FAVOURITE TREE IS ACACIA, ZIZIPHUS OR BIRCH
IF YOU'VE TAKEN THE DRONGO'S SPOT, HE WILL NOT LET YOU PERCH!
NO THREAT IS A THREAT BIG ENOUGH FOR THE DRONGO, BE IT A CROCODILE OR CAIMAN,
SNAP!
OR A CROW OR A CAT, OR AN EAGLE OR A KITE, OR THE GREAT GENGHIS KHAN.
IT IS HENCE NO SURPRISE THAT OTHER NESTING BIRDS, AROUND THE DRONGO'S NEST GATHER,
AND KISS HIS HAND IN DEFERENCE, SAYING, 'MUAH, GODFATHER!'

#22
SUSTAINABLE TRAVEL TIPS FROM THE ARCTIC TERN
1. ALWAYS FLY ECONOMY CLASS.
2. EAT LOCAL.
Antarctic Herring
3. DO NOT LITTER.
munch munch
pthoo
4. MINIMIZE YOUR WATER USAGE.
5. ALWAYS ASK BEFORE YOU TAKE PICTURES OF LOCALS OR THEIR RITUALS.
MAY I?
6. USE PUBLIC TRANSPORT.
DOES THIS GO TO VOSTOK?
7. BUY LOCALLY PRODUCED GOODS.
FOUR OF THOSE, PLEASE.
ANTARCTIC HERRING BONE COMB
SOUVENIR SHOP
8. DO NOT DISTURB NATIVE WILDLIFE. KEEP A SAFE DISTANCE.
9. CHOOSE NON-POLLUTING RECREATIONAL ACTIVITIES.
WHEEEE!

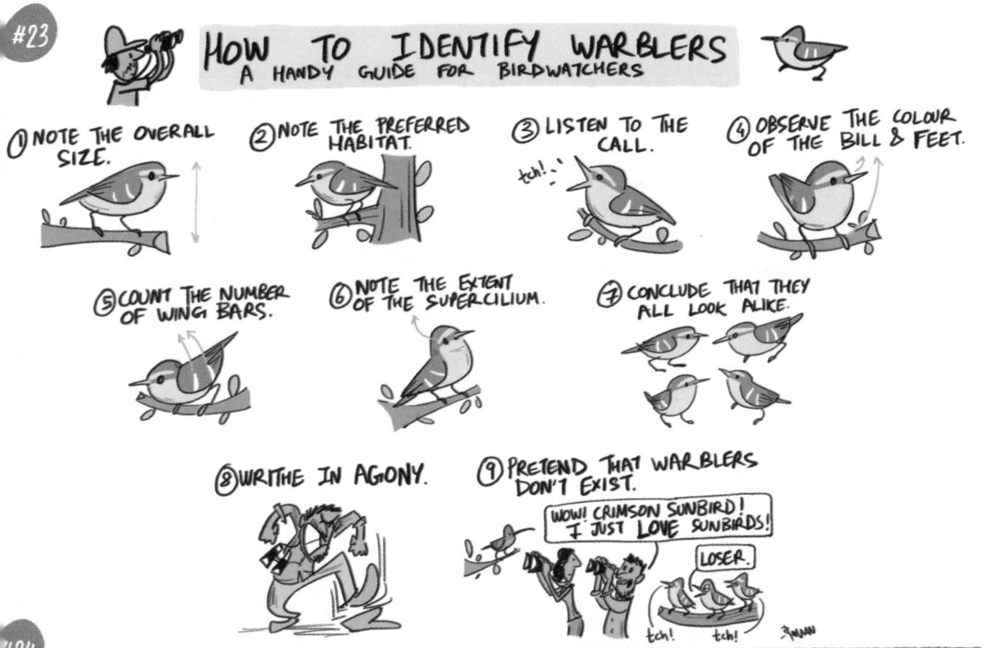
#23
HOW TO IDENTIFY WARBLERS
A HANDY GUIDE FOR BIRDWATCHERS
① NOTE THE OVERALL SIZE.
② NOTE THE PREFERRED HABITAT.
③ LISTEN TO THE CALL.
tch!
④ OBSERVE THE COLOUR OF THE BILL & FEET.
⑤ COUNT THE NUMBER OF WING BARS.
⑥ NOTE THE EXTENT OF THE SUPERCILIUM.
⑦ CONCLUDE THAT THEY ALL LOOK ALIKE.
⑧ WRITHE IN AGONY.
⑨ PRETEND THAT WARBLERS DON'T EXIST.
WOW! CRIMSON SUNBIRD! I JUST LOVE SUNBIRDS!
LOSER.
tch!
tch!

#24
THIS IS MY NEIGHBOURHOOD IN CENTRAL INDIA.
THESE ARE THE CAUCASUS MOUNTAINS OF WEST ASIA.
GUESS WHAT THE CONNECTING LINK BETWEEN THESE TWO PLACES IS?
THIS TINY, DRAB-LOOKING BIRD THAT VISITS THE BACKYARD TREE EVERY WINTER: THE GREENISH WARBLER!
chilip!
RESEARCH HAS ESTABLISHED THAT WARBLERS VISITING INDIA RETURN TO THE EXACT SAME TERRITORY YEAR AFTER YEAR, EACH WINTER!
A 7-GRAM BIRD FLYING 5000 KM, CONNECTING TWO DISTANT LANDS, AND TEACHING US THE SIGNIFICANCE OF EVERY SINGLE NEIGHBOURHOOD TREE: ISN'T MIGRATION A TRULY WONDERFUL PHENOMENON?
INDEED! LAST WEEK I WAS DIPPING MY WORMS IN HUMMUS. TODAY IT'S MANGO PICKLE!

#25

FOOD CAN BE VERY SCARCE IN WINTERS IN THE HIMALAYAS. SO WE NUTCRACKERS HAVE AN INGENIOUS SOLUTION—
WE CACHE FOOD LIKE PINE CONES BY BURYING THEM UNDERGROUND OR WEDGING THEM BETWEEN ROCKS USING OUR DAGGER-SHAPED BILLS.
OUR INCREDIBLE SPATIAL MEMORY GUIDES US TO MOST OF OUR CACHES IN WINTER, ENSURING WE DON'T STARVE...
BUT WHAT HAPPENS TO THE CACHES WE FORGET? YOU GUESSED IT— THEY GROW INTO PINE FORESTS!
AMONG THESE ARE TREES LIKE THE CHILGOZA PINE – A NUTRITIOUS & COMMERCIALLY IMPORTANT PINE FACING A THREAT OF EXTINCTION FROM OVER-HARVESTING, HABITAT LOSS & CLIMATE CHANGE.
AND THIS ABILITY TO DISPERSE PINES MAKES US NUTCRACKERS CRUCIAL FARMERS OF TEMPERATE FORESTS!
FORESTRY ISN'T A TOUGH NUT TO CRACK IF YOU CONSULT THE RIGHT EXPERTS.

#26

THE HORNBILL DAD

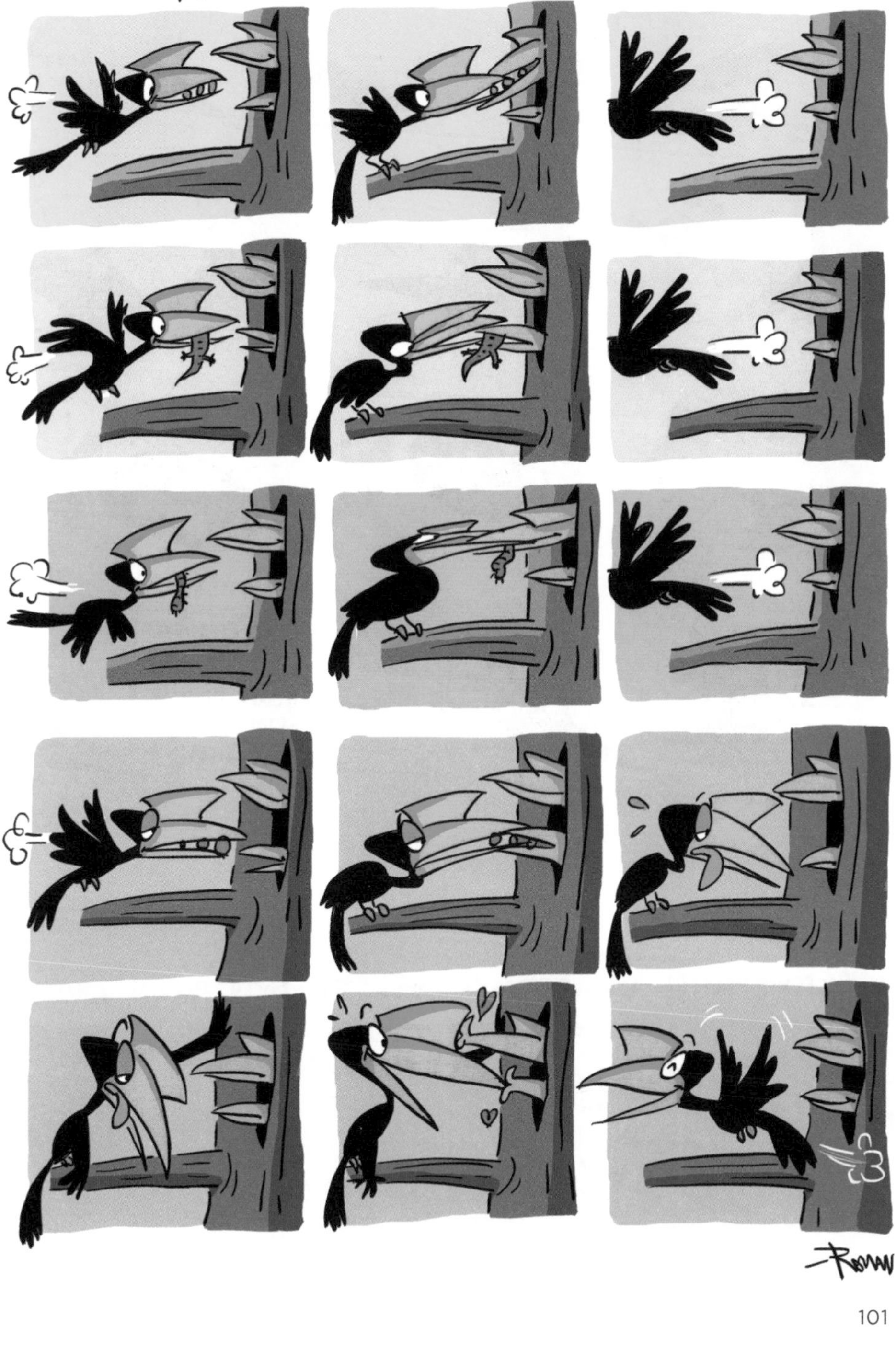
ROMAN

#27

THE COURTSHIP DANCE OF THE WESTERN PAROTIA

SWEEP SWEEP
DUST DUST
SHAKE SHAKE
!
SO, WHICH OF MY MOVES DID YOU LIKE THE BEST?
THE EFFICIENCY WITH WHICH YOU CLEANED THE FLOOR. NOTHING LIKE A GUY WHO KNOWS HIS HOUSEKEEPING.

#28

A VOGELKOP SUPERB BIRD-OF-PARADISE TAKES THE DOLLY PARTON CHALLENGE

LINKEDIN
FACEBOOK
INSTAGRAM
TINDER

#29

TOP TEN RECOMMENDATIONS FOR MIGRATORY BIRDS VISITING INDIA
JET-LAGGED? TAKE A WARM DUST BATH IN RED SOIL.
TREAT YOURSELF TO SOME FAT, JUICY MICE IN OUR FIELDS!
PAMPER YOUR PALATE WITH OUR DELECTABLE COASTAL CUISINE.
INDULGE IN SOME LOCALLY BREWED MAHUA LIQUOR.
SAMPLE SOME SUSTAINABLE COFFEE FROM THE WESTERN GHATS. (PREFERRED BRAND: BLACK BAZA)
GO MONUMENT-HOPPING WITH OUR LOCAL GUIDES.
WITNESS THE BEAUTY OF INDIAN CLASSICAL DANCE.
BREAK THE LAW! TRY SOME COUNTRY BEEF.
IF YOU'RE A VISITING CUCKOO, DROP YOUR 'LUGGAGE' AT A NANNY'S NEST!
HANG OUT WITH COUCHSURFERS FROM INDIA AND ABROAD!

#30

HOW VAGRANT MIGRATIONS HAPPEN:

GOODNESS GRACIOUS ME !! A BLACK-LEGGED KITTIWAKE! A SEA BIRD STRAYING RIGHT INTO THE MIDDLE OF INDIA! WE'RE GOING TO MAKE HISTORY!!
CRASH!
@#%*ING APPLE MAPS.

#31

LOSE 50Kg! IN 2 WEEKS!
ENROLL NOW!
IF THEY TRY MIGRATING NON-STOP FOR 2000 KILOMETRES ON LIMITED RESERVES, THEY'LL KNOW THE REAL VALUE OF BODY FAT.

32

PARROW SONGS FROM AROUND THE WORLD
♪ chweep ♫
HOUSE SPARROW
♪ el chweep ♫
SPANISH SPARROW
@#*%.ING CHWEEP!
ROCK SPARROW
♪ AL-chweep ♫
ARABIAN SPARROW
♪ la-chweep ♫
ITALIAN SPARROW
song.open( );
song.sing("<c1>CHWEEP</c1>");
song.close( );
JAVA SPARROW

WHO NEEDS IKEA!

33

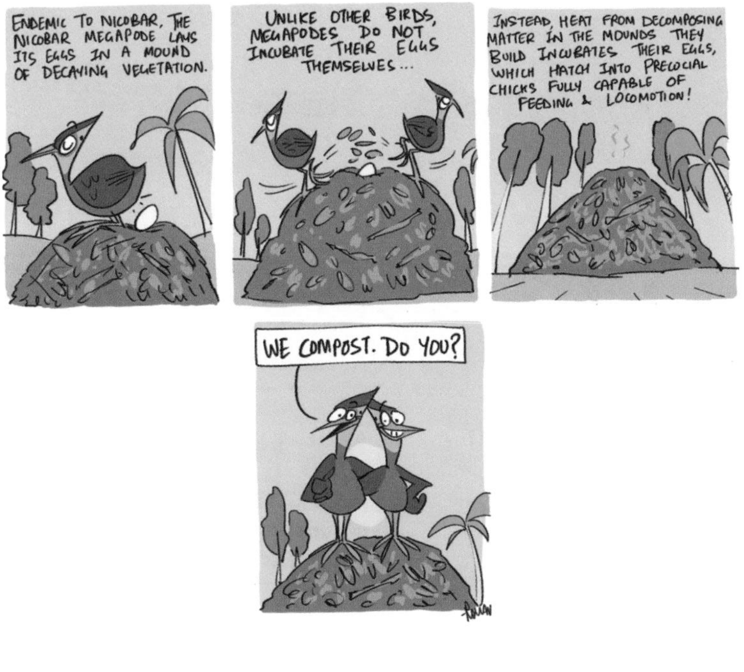
ENDEMIC TO NICOBAR, THE NICOBAR MEGAPODE LAYS ITS EGGS IN A MOUND OF DECAYING VEGETATION.
UNLIKE OTHER BIRDS, MEGAPODES DO NOT INCUBATE THEIR EGGS THEMSELVES...
INSTEAD, HEAT FROM DECOMPOSING MATTER IN THE MOUNDS THEY BUILD INCUBATES THEIR EGGS, WHICH HATCH INTO PRECOCIAL CHICKS FULLY CAPABLE OF FEEDING & LOCOMOTION!
WE COMPOST. DO YOU?

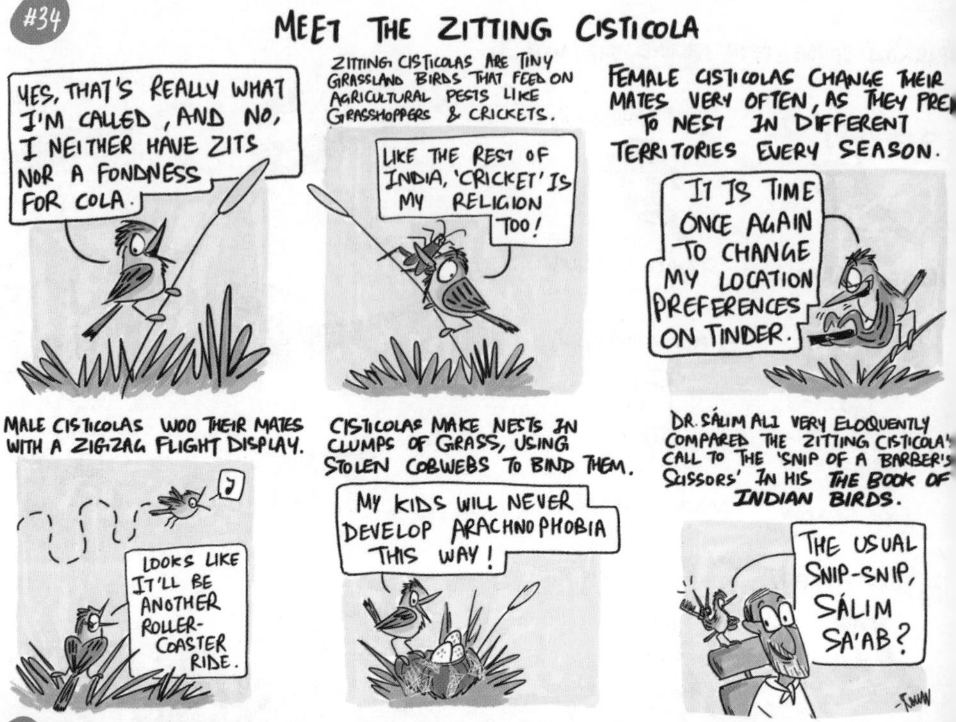
#34
MEET THE ZITTING CISTICOLA
YES, THAT'S REALLY WHAT I'M CALLED, AND NO, I NEITHER HAVE ZITS NOR A FONDNESS FOR COLA.
ZITTING CISTICOLAS ARE TINY GRASSLAND BIRDS THAT FEED ON AGRICULTURAL PESTS LIKE GRASSHOPPERS & CRICKETS.
LIKE THE REST OF INDIA, 'CRICKET' IS MY RELIGION TOO!
FEMALE CISTICOLAS CHANGE THEIR MATES VERY OFTEN, AS THEY PRE TO NEST IN DIFFERENT TERRITORIES EVERY SEASON.
IT IS TIME ONCE AGAIN TO CHANGE MY LOCATION PREFERENCES ON TINDER.
MALE CISTICOLAS WOO THEIR MATES WITH A ZIGZAG FLIGHT DISPLAY.
LOOKS LIKE IT'LL BE ANOTHER ROLLER-COASTER RIDE.
CISTICOLAS MAKE NESTS IN CLUMPS OF GRASS, USING STOLEN COBWEBS TO BIND THEM.
MY KIDS WILL NEVER DEVELOP ARACHNOPHOBIA THIS WAY!
DR. SÁLIM ALI VERY ELOQUENTLY COMPARED THE ZITTING CISTICOLA' CALL TO THE 'SNIP OF A BARBER'S SCISSORS' IN HIS THE BOOK OF INDIAN BIRDS.
THE USUAL SNIP-SNIP, SÁLIM SA'AB?

#35
Sniff sniff
sniff sniff
sniff
dig dig
sniff sniff
sniff
sniff sniff
DAMN IT! I CAN SMELL MY PREY BUT JUST CAN'T SPO...
Whrrrr!
APRIL IS A NIGHTJAR'S FAVOURITE MONTH!

#36

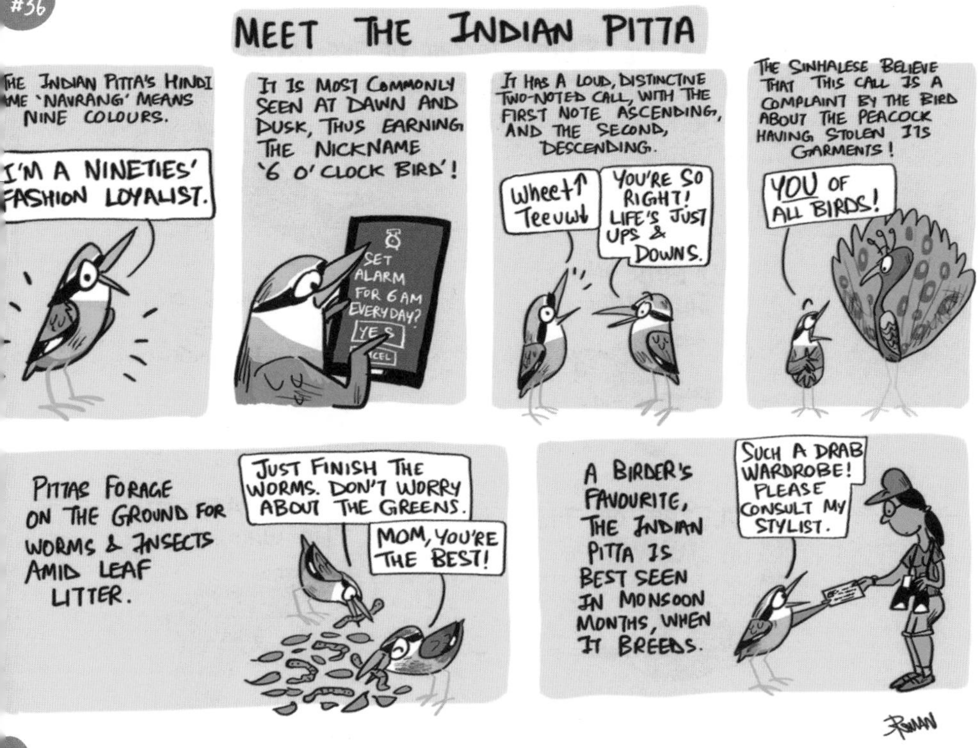

MEET THE INDIAN PITTA
THE INDIAN PITTA'S HINDI
ME 'NAVRANG' MEANS NINE COLOURS.
I'M A NINETIES' FASHION LOYALIST.
IT IS MOST COMMONLY SEEN AT DAWN AND DUSK, THUS EARNING THE NICKNAME '6 O' CLOCK BIRD'!
SET ALARM FOR 6 AM EVERYDAY?
YES
IT HAS A LOUD, DISTINCTIVE TWO-NOTED CALL, WITH THE FIRST NOTE ASCENDING, AND THE SECOND, DESCENDING.
Wheet↑ Teeuw↓
YOU'RE SO RIGHT! LIFE'S JUST UPS & DOWNS.
THE SINHALESE BELIEVE THAT THIS CALL IS A COMPLAINT BY THE BIRD ABOUT THE PEACOCK HAVING STOLEN ITS GARMENTS!
YOU OF ALL BIRDS!
PITTAS FORAGE ON THE GROUND FOR WORMS & INSECTS AMID LEAF LITTER.
JUST FINISH THE WORMS. DON'T WORRY ABOUT THE GREENS.
MOM, YOU'RE THE BEST!
A BIRDER'S FAVOURITE, THE INDIAN PITTA IS BEST SEEN IN MONSOON MONTHS, WHEN IT BREEDS.
SUCH A DRAB WARDROBE! PLEASE CONSULT MY STYLIST.
ROHAN

37

WORLD RECORDS BY MIGRATORY BIRDS
LONGEST MIGRATION
ARCTIC TERN
HIGHEST MIGRATION FLIGHT
BAR-HEADED GOOSE
FASTEST MIGRATION FLIGHT
GREAT SNIPE
LONGEST NON-STOP MIGRATION
BAR-TAILED GODWIT
LONGEST MIGRATION BY A SONGBIRD
NORTHERN WHEATEAR
HIGHEST NUMBER OF INSTAGRAM FOLLOWERS
RUFF

#38

THE SWIFT IS ONE OF THE FASTEST AND THE MOST AERIAL OF ALL BIRDS!
IT EATS ON THE WING...
DRINKS ON THE WING...
SLEEPS ON THE WING...
AND EVEN MATES ON THE WING!
ALL OF THIS WITHOUT ANY REDBULL!

#39

LADIES AND GENTLEMEN, I PRESENT TO YOU INDIA'S RESIDENT BIRD WITH THE LARGEST WINGSPAN- THE HIMALAYAN GRIFFON!
SO HOW WIDE IS YOUR WINGSPAN REALLY, MA'AM?
OH, VERY, VERY WIDE!
ABOUT AS WIDE AS THE DIFFERENC
BETWEEN PATRIOTISM FOR A POLITICAL PARTY AND PATRIOTISM FOR A NATION.

#40
DAD WITH A DIFFERENCE - THE PHALAROPE FATHER
A COMPLETE GENDER BENDER!
MY WIFE WEARS THE BREEDING PLUMAGE IN THIS HOUSEHOLD!
SOLELY RESPONSIBLE FOR ALL HOUSEHOLD CHORES.
NEVER TOO BUSY TO TAKE THE KIDS FOR A SWIM.
FEIGNS AN INJURY & RISKS HIS OWN LIFE TO DISTRACT PREDATORS AWAY FROM HIS BROOD.
DOESN'T JUDGE HIS MATE FOR HER UNCONVENTIONAL LIFESTYLE.
HONEY, I'M ON A TINDER DATE. I THINK I LIKE HIM & I'M GOING TO START A NEW FAMILY.
GOOD LUCK, SWEETHEART. DON'T WORRY ABOUT THE KIDS.

#41
LEARN PATRIOTISM FROM THE CATTLE EGRET
① SHARES A SYMBIOTIC RELATIONSHIP WITH COWS WITHOUT JUDGING OTHERS WHO DON'T.
② NESTS PEACEFULLY ALONGSIDE FELLOW COUNTRYMEN OF A DIFFERENT CREED AND COLOUR.
③ COMMITTED TO SERVING OFFICE DESPITE INHERENT NARCISSISM.
④ PROUDLY WEARS SAFFRON WITHOUT GOING ON A RAMPAGE.

#42

HEY! AREN'T YOU THE SATELLITE-TAGGED CUCKOO THAT REACHED YEMEN FROM INDIA IN JUST TWO DAYS, SETTING A NEW MIGRATION RECORD?! HOW DID YOU ACCOMPLISH THIS FEAT?
WELL, I GUESS NOT HAVING TO RAISE MY OWN KIDS HELPS ME FOCUS ON MY CAREER GOALS BETTER.
THAT'S SUCH A DISGUSTING THING
SAY! ARCTIC TERNS ARE DEDICATED PARENTS AND YET THEY FLY FARTHER THAN YOU, YOU SELFISH $#%@!
WHAT DO YOU CUCKOOS KNOW ABOUT THE BEAUTY OF RAISING A CHICK? HMPH. NOW EXCUSE ME; I CANNOT BEAR THE SIGHT OF SUCH A TRASHY CREATURE.
AND THAT WAS MY TUTORIAL FOR BUDDING BROOD PARASITES ON HOW TO PISS OFF NESTING BIRDS.
PLOP!
ROHAN

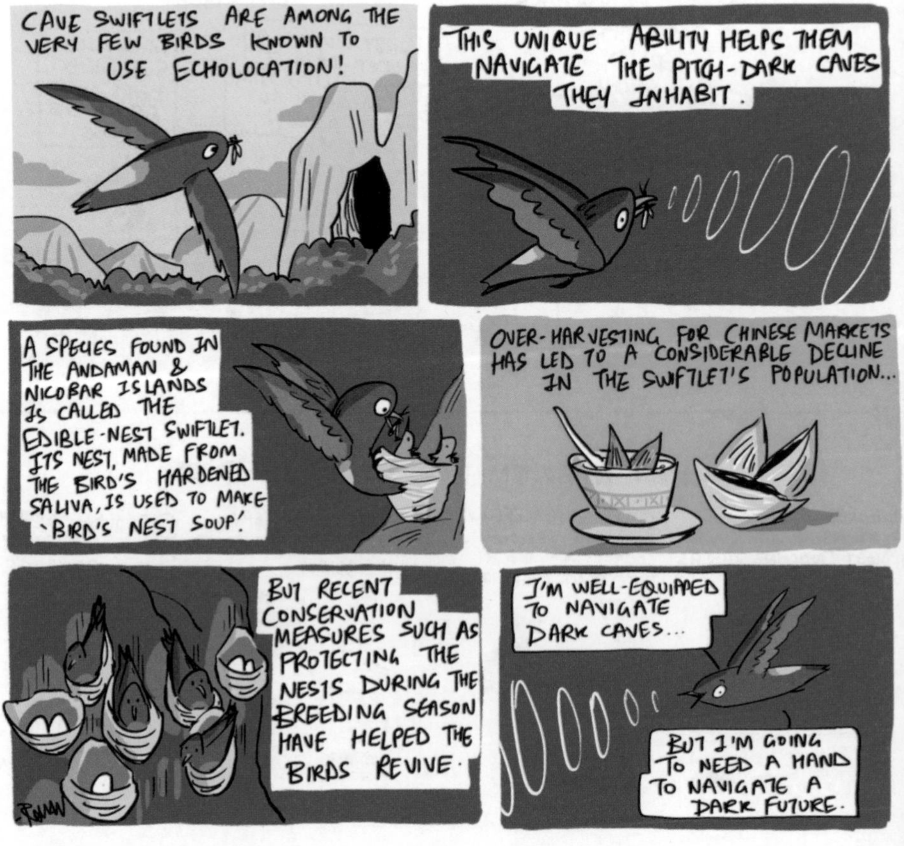
CAVE SWIFTLETS ARE AMONG THE VERY FEW BIRDS KNOWN TO USE ECHOLOCATION!
THIS UNIQUE ABILITY HELPS THEM NAVIGATE THE PITCH-DARK CAVES THEY INHABIT.
A SPECIES FOUND IN THE ANDAMAN & NICOBAR ISLANDS IS CALLED THE EDIBLE-NEST SWIFTLET. ITS NEST, MADE FROM THE BIRD'S HARDENED SALIVA, IS USED TO MAKE 'BIRD'S NEST SOUP'.
OVER-HARVESTING FOR CHINESE MARKETS HAS LED TO A CONSIDERABLE DECLINE IN THE SWIFTLET'S POPULATION...
BUT RECENT CONSERVATION MEASURES SUCH AS PROTECTING THE NESTS DURING THE BREEDING SEASON HAVE HELPED THE BIRDS REVIVE.
ROHAN
I'M WELL-EQUIPPED TO NAVIGATE DARK CAVES...
BUT I'M GOING TO NEED A HAND TO NAVIGATE A DARK FUTURE.

#43

FIVE
Reptiles, Amphibians and **Invertebrates**

CONSERVATION THREATS TO FROGS
•Climate change
•Water pollution
•Chytridiomycosis
• Pesticides
• Habitat loss
• Wetland encroachment
• Deforestation
• Invasive species
• Over-harvesting

#1

NO, THANKS. I ALREADY HAVE TOO MUCH TO DEAL WITH.
ROMAN

Our unanimous favourites may be the more glamorous feathered and furred stars of the animal kingdom, but it is the neglected supporting cast that keeps the nuts and bolts of our planet oiled and functioning: the reptiles, amphibians and invertebrates.

Here are some creepy-crawlies creeping and crawling about in comics. I promise there will be no more alliterations in this chapter (okay, maybe just some more).

#2

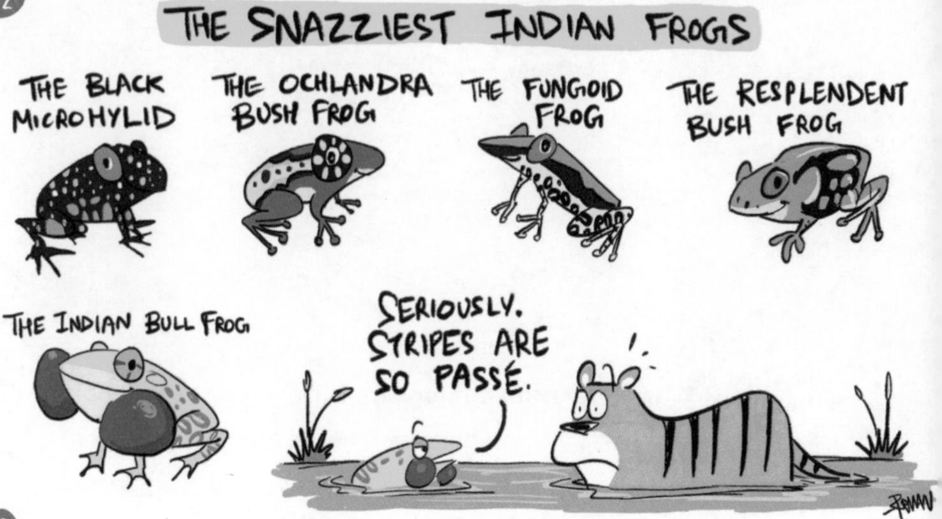
THE SNAZZIEST INDIAN FROGS
THE BLACK MICROHYLID
THE OCHLANDRA BUSH FROG
THE FUNGOID FROG
THE RESPLENDENT BUSH FROG
THE INDIAN BULL FROG
SERIOUSLY. STRIPES ARE SO PASSÉ.

#3

YOU'RE RUDE TO MY FRIENDS, YOU CALL MY MOTHER A COW, YOU'RE A JUDGEMENTAL PRICK & AN INCORRIGIBLE JERK. GIVE ME ONE GOOD REASON TO CARRY ON WITH YOU.
SEXY B**TARD.

SNAP!
12
1
2
3
CATCH!
NIP!
CRUNCH!
CHUCK, CHUCK, CHUCK, CHUCK, CHUCK!*
* YOU'LL BE PLEASED TO KNOW THAT YOUR ROOM HAS BEEN CLEARED OF ALL BUGS & PESTS FOR THE NIGHT, SIR!
WAAAAA!
CHUCK, CHUCK, CHUCK, CHUCK.*
* A 'THANK YOU' WOULD'VE BEEN NICE.

#5

HERE'S A LIST OF CONSERVATION THREATS FACED BY BEES-
PESTICIDES & CHEMICALS, STRESS FROM CLIMATE CHANGE, LOSS OF HABITAT, GENETICALLY MODIFIED CROPS, CELLPHONE RADIATION...
WHY SHOULD THIS BOTHER YOU, YOU MIGHT ASK. TWO REASONS-
FIRST- WE POLLINATE JUST ABOUT EVERY CROP YOU DEPEND ON.
AND SECOND, WHEN WE GO EXTINCT, WE'RE TAKING YOU WITH US!
RESEARCH PREDICTS THAT HUMANS WILL BE EXTINCT WITHIN YEARS OF THE DISAPPEARANCE OF BEES FROM THE PLANET.
DON'T BE SCARED. EXTINCTION IS THE LATEST 'BUZZ WORD'!

#6

SOME SIMILARITIES BETWEEN THE APATANI GLORY MOTH AND THE APATANI PEOPLE OF ARUNACHAL

INCREDIBLY ATTRACTIVE

VERY FOND OF DESIGNER WEAR

MAKE WONDERFUL USE OF NATURAL RESOURCES

HAVE THEIR OWN BRAND OF JEWELLERY & BODY ART

GRACEFUL DANCE MOVES

LOVE THEIR NECTAR!

"POREY-O" OR RICE BEER.
-RuMAN

#7

THINGS MOTHS CAN DO, BUT I CAN'T

SUSTAIN AN ENTIRE ECOSYSTEM

SIZZLE IN GAUDY CLOTHING

SIZZLE IN DRAB CLOTHING

PRETEND TO BE BIRD POOP TO AVOID UNWANTED COMPANY

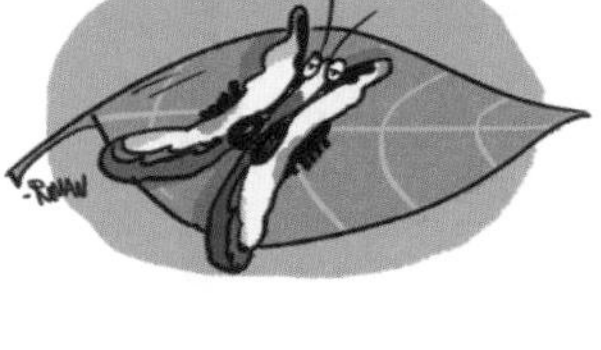
-RuMAN

NOT BE ASHAMED OF TEENAGE PICTURES

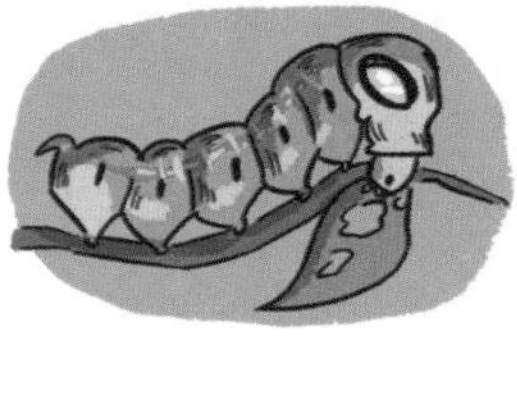

ACTUALLY APPLY MATHEMATICS TO LIFE

deflected bat sonar

#8

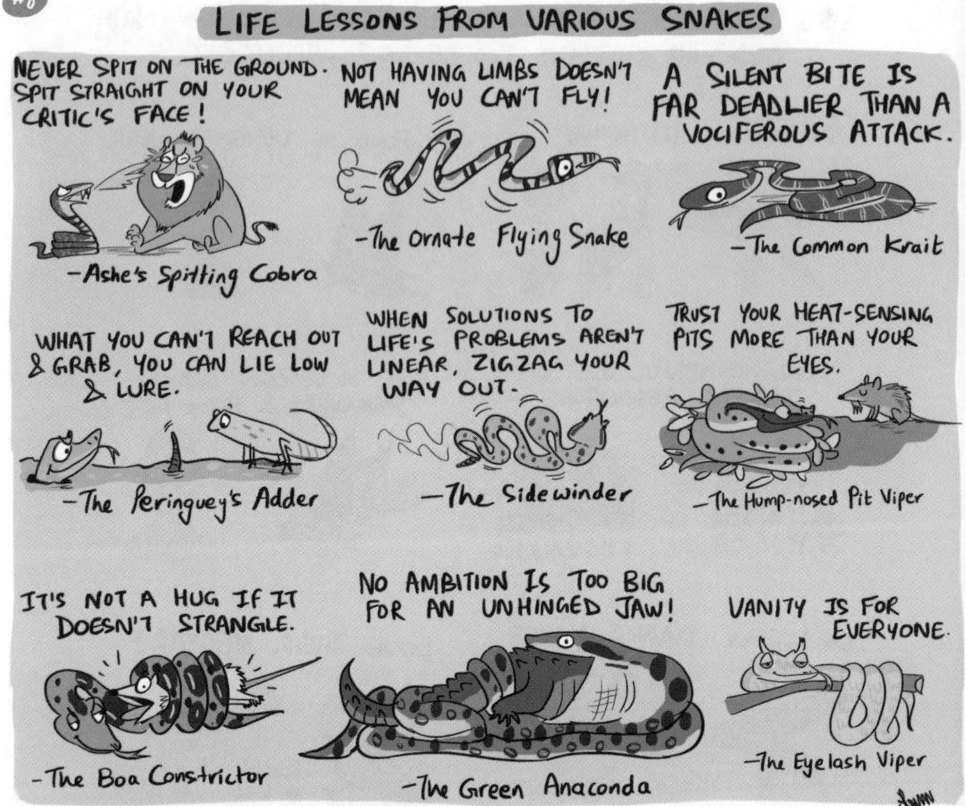
LIFE LESSONS FROM VARIOUS SNAKES
NEVER SPIT ON THE GROUND. SPIT STRAIGHT ON YOUR CRITIC'S FACE!
-Ashe's Spitting Cobra
NOT HAVING LIMBS DOESN'T MEAN YOU CAN'T FLY!
-The Ornate Flying Snake
A SILENT BITE IS FAR DEADLIER THAN A VOCIFEROUS ATTACK.
-The Common Krait
WHAT YOU CAN'T REACH OUT & GRAB, YOU CAN LIE LOW & LURE.
-The Peringuey's Adder
WHEN SOLUTIONS TO LIFE'S PROBLEMS AREN'T LINEAR, ZIGZAG YOUR WAY OUT.
-The Sidewinder
TRUST YOUR HEAT-SENSING PITS MORE THAN YOUR EYES.
-The Hump-nosed Pit Viper
IT'S NOT A HUG IF IT DOESN'T STRANGLE.
-The Boa Constrictor
NO AMBITION IS TOO BIG FOR AN UNHINGED JAW!
-The Green Anaconda
VANITY IS FOR EVERYONE.
-The Eyelash Viper

#9

SCRATCH
#makeover #newyearnewme

#10

SOME OF INDIA'S BEST DANCERS HAIL FROM THE SOUTH. ONE OF THEM IS THE INDIAN DANCING FROG FROM THE WESTERN GHATS!
LIVING IN FAST-FLOWING FOREST STREAMS MEANS THAT YOUR CROAKS AREN'T ALWAYS AUDIBLE TO A POTENTIAL MATE...
TO OVERCOME THIS, THE MALE DANCING FROG EXTENDS HIS HIND LEG & HOLDS IT OUT, PRESENTING HIS FLASHY WEBS TO THE LADIES!
THE MOVE IS ALSO USED TO KICK RIVAL MALES OFF THEIR PERCHES!
HERPETOLOGISTS CALL THIS BEHAVIOUR 'FOOT-FLAGGING'...
BUT I CALL IT A FUSION OF BHARATANATYAM & KALARIPAYATTU!

#11
HOW TO OVERCOME A CREATIVE BLOCK:
A GUIDE BY THE COMMON BLUEBOTTLE BUTTERFLY.
BE KIND TO YOURSELF. BINGE VORACIOUSLY ON YOUR FAVOURITE JUNK FOOD.
FIND A QUIET CORNER & PUPATE.
REMEMBER: KEEPING AWAY FROM SOCIAL MEDIA IS IMPERATIVE TO METAMORPHOSIS.
BORED OF MEDITATING? BREAK FREE; EXERCISE YOUR WINGS.
GO GRAB A QUICK DRINK.
GO MUD-PUDDLING WITH FELLOW ARTISTS.
MAKE LOVE.
JUST SIT & WATCH THIS COLOURFUL WORLD GO BY, UNTIL THE DEADLINE PANIC STRIKES.
NOW GO MAKE SOME ART!

#12
SOME TALENTS OF THE DRACO I WISH I HAD
GLIDE AT WILL
Wheee!
AVOID UNWANTED COMPANY USING CAMOUFLAGE
ROCK A BRIGHT YELLOW TIE
TUMBLE WITH THE GRACE OF A FALLING LEAF
BEAT DC & MARVEL SUPERHEROES AT COSPLAY

#13
I'VE DONE IT! I'VE SCALED MOUNT EVEREST!
CLAP CLAP CLAP!
DON'T MEAN TO BURST YOUR BUBBLE, BUT I GOT HERE LONG BEFORE YOU DID. MY SPECIES HAS LIVED ON THE EVEREST FOR MILLIONS OF YEARS, WITHOUT ANY MOUNTAINEERING GEAR!
I'M THE HIMALAYAN JUMPING SPIDER, THE HIGHEST LIVING ORGANISM ON PLANET EARTH.
BUT TO BE FAIR, CONGRATULATIONS ON YOUR ACHIEVEMENT, SON. HERE'S A SOUVENIR FOR YOU...
A SIGNED SELFIE OF MY GREAT GRANDPA WITH HIS BUDDIES, EDMUND & TENZING.

#14

THE SPIDER-WASP MOTHER

THE SPIDER-WASP MOTHER IS LIKE NO OTHER,
AS DEVOTED A PARENT AS CAN BE,
ARMED WITH A STING THAT'S PA-RA-LY-SING,
SHE'D KILL FOR HER YOUNG WITH GLEE.
AS THE NAME INDICATES, HER PREY'S LEGGED EIGHT:
A TARANTULA TODAY SHE STALKS,
A NASTY FIGHT; A STING AND A BITE,
AND THE SPIDER'S LEFT STUNNED AS A ROCK!
SHE CARRIES OFF HER QUARRY; SHE'S GOT TO HURRY
LEST PIRATES & ROBBERS APPEAR,
NOTWITHSTANDING THAT THE BIG, BROWN THING
WEIGHS THREE TIMES HER OWN DERRIÈRE!
IN CALCULATED HASTE, SHE DIGS UP A NEST
AND DUMPS HER CATCH WITH A THUD!
SHE THEN LAYS HER EGG ON MISTER HAIRY-VON-GREG
AND BURIES HIM BACK IN THE MUD.
WHEN THE LARVA COMES OUT, HE HAS NOTHING TO FRET ABOUT,
ALL THE BREAD'S HIS TO BREAK,
'CAUSH HIS SPIDER-WASP MOTHER, WHO'S LIKE NO OTHER,
BAKED HIM A GIANT BIRTHDAY CAKE.
BURP!
-RWW

#15
SOME WORLD RECORDS HELD BY THE GLOBE SKIMMER DRAGONFLY
ITS 6000 km-LONG MIGRATION IS THE LONGEST FOR ANY INSECT ON EARTH!
GLOBE SKIMMER is travelling from TOKYO INTERNATIONAL AIRPORT to CHENNAI
SO JEALOUS!!!
TAKE ME ALONG!
HOW DO YOU EVEN ??!!
IT IS THE MOST WIDESPREAD DRAGONFLY SPECIES, FOUND ON EVERY CONTINENT EXCEPT ANTARCTICA.
I CAN GIVE YOU COUCHSURFING RECOMMENDATIONS FROM AROUND THE GLOBE!
IT IS THE ONLY SPECIES OF DRAGONFLY FOUND ON EASTER ISLAND.
CLINK
CLINK
IT IS THE FIRST SPECIES OF DRAGONFLY TO SETTLE IN BIKINI ATOLL AFTER NUCLEAR TESTS WERE CONDUCTED THERE.
I THINK THE USA DESERVES DONALD TRUMP FOR WHAT IT DID TO BIKINI.
GLOBE SKIMMERS HAVE BEEN RECORDED FLYING AT 6200 m IN THE HIMALAYAS, MAKING THEM THE HIGHEST FLYING INSECTS!
GLOBE SKIMMER EVEREST WATCHING TOURS. BOOK NOW!

#16
WHAT WE DO TO DAMSELFLIES:
-DESTRUCTION OF THEIR HABITAT: FOREST STREAMS, WETLANDS & MARSHES.
-DEFORESTATION.
-DAMS, MINES & UNPLANNED DEVELOPMENT.
-CLIMATE CHANGE & IRREGULAR RAINFALL.
WHAT DAMSELFLIES DO FOR US:
-PLEASE OUR EYES WITH THEIR ETHEREAL BEAUTY.
-PREY ON MOSQUITOES & PESTS, SAVING US FROM DISEASE & FAMINE.
GREAT. I'M THE DAMSEL IN DISTRESS, AND YET I DO ALL THE SAVING!

#17

A GIANT WOODSPIDER GOES ON A TINDER DATE

LOOK, BEFORE WE BEGIN, I'D LIKE TO MAKE ONE THING VERY CLEAR.
MMHMM?
I'M NOT THE KIND OF WOMAN WHO'S JUST LOOKING FOR SEX ON TINDER.
I MEAN I HAVE NOTHING AGAINST THAT, BUT I PERSONALLY AM SEEKING SOMETHING MORE MEANING-FUL.
LIKE WHAT?
LIKE FOOD.
ROMAN

#18

'META' MORPHOSIS

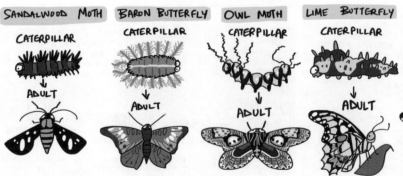

SANDALWOOD MOTH
CATERPILLAR
ADULT
BARON BUTTERFLY
CATERPILLAR
ADULT
OWL MOTH
CATERPILLAR
ADULT
LIME BUTTERFLY
CATERPILLAR
ADULT
REST ASSURED KIDS. THE WEIRDEST TEENAGERS GR UP INTO THE SEXIEST ADUL

#19

HOW DUNG BEETLES ROLL

DUNG BEETLES EVOLVED MILLIONS OF YEARS AGO WHEN TRULY ENORMOUS MAMMALS ROAMED THE EARTH.
SPLAT!
GOOD TIMES!
DUNG BEETLES DERIVE ALL THEIR NUTRITION & MOISTURE FROM DUNG, AND THEY DON'T NEED TO DRINK!
IT'S A 1985 CHATEAU DÉ ELEPHANTÉ.
DUNG BEETLES DO A LOT MORE THAN EATING DUNG. THEY RECYCLE NUTRIENTS BY BREAKING THE DUNG UP, IMPROVING SOIL QUALITY.
DUNG BEETLE SCHOOL OF AGRICULTURE AND LAND MANAGEMENT
MALE DUNG BEETLES USE DUNG TO WOO FEMALES.
PROMISE ME YOU'LL ALWAYS TREAT ME LIKE SHIT.
PROMISE.
DUNG BEETLES ARE AMONG THE WORLD'S STRONGEST CREATURES, ABLE TO PUSH UP TO TEN TIMES THEIR BODY WEIGHT!
SISYPHUS, YOU SISSY!
DUNG BEETLES ALSO DISPERSE SEEDS, PLAYING AN IMPORTANT ROLE IN FOREST REGENERATION.
WHEN LIFE GIVES ME SHIT, I TURN IT INTO A FOREST.
DUNG BEETLES PROTECT CATTLE FROM PESTS SUCH AS HORNFLIES, BY REMOVING THE DUNG THAT HOSTS THESE PESTS.
THANK YOU.
NO, THANK YOU!
THAT'S GREAT-GREAT-GREAT-GREAT-GREAT-GREAT GRANDPA, ADVISOR TO RAMSES VI.
'SUP?
NOTHING MUCH. JUST KEEPING THE PLANET ROLLING.
-ROHAN

#20
SOME SIMILARITIES BETWEEN DAVID ATTENBOROUGH AND THE ATTENBOROUGH'S FLAT LIZARD
MARKED AGILITY
A SPARKLING SCREEN PRESENCE
IMPECCABLE DRESSING
CHICK MAGNETS

#21
I'M THE PARASITIC LARVA OF AN ICHNEUMON WASP, AND THIS IS MY HOST, AN ORB-WEAVING SPIDER.
MY MOTHER STUNG THIS BEAUTY, AND PLANTED ME IN HER BODY. SINCE THEN, I'VE BEEN EATING HER ALIVE...
THE POOR SPIDER GOES ABOUT HER BUSINESS, WEAVING HER LOVELY WEB AND CATCHING HER PREY, WITHOUT REALIZING THAT I REAP THE BENEFITS OF ALL HER LABOUR!
BEFORE I SUCK THE LAST BREATH OUT OF HER BODY, I WILL MANIPULATE HER INTO WEAVING HER FINAL CREATION— A THREAD OF STURDY SILK CABLES TO PROTECT MY COCOON, SECURING MY FUTURE EVEN AFTER SHE IS DEAD!
WOW.
YOU KNOW WHAT?
YOU SHOULD BE IN PUBLISHING.

#22
THINGS I NEED TO LEARN FROM MILLIPEDES
MOVING AT YOUR OWN PACE, EVEN IF YOU HAVE HUNDREDS OF LEGS.
KEEPING AWAY FROM THE LIMELIGHT DESPITE PERFORMING ONE OF THE MOST IMPORTANT TASKS FOR THE PLANET: SOIL FORMATION AND NUTRIENT RECYCLING.
TO NOT JUST BE AN ARTIST BUT TO ADD VALUE TO AN ART FORM (MILLI COMPOST IS RICHER THAN VERMICOMPOST!).
EMBRACING CHANGE, GROWING NEW LEGS & SEGMENTS ON EVERY MOULT.
ROLLING INTO A DEFENSIVE COIL WHEN PREDATORS OR TROLLS APPROACH.
MORE 'ME TIME.'

#23
MONSOON MELODIES BY INDIAN FROGS
KREEA-KROOAA-KREEA-KROOAA
INDIAN BULL FROG
tikitikitikitiktik!
COMMON SKITTERING FROG
BRRUMP! BRRUMP!
PAINTED KALOULA
titik-titik-titik
KOTTIGEHARA DANCING FROG
PAINP! PAINP! PAINP!
VARIEGATED RAMANELLA
KROUUWA
JOG NIGHT FRO
TRREE-tink-tink-tink
YELLOW BUSH FROG
PIHEEUWW-pip-pip-pip!
KEMPHOLEY NIGHT FROG
MODERN BOLLYWOOD MUSIC 'CROAKS'. NOT FROGS.

#24

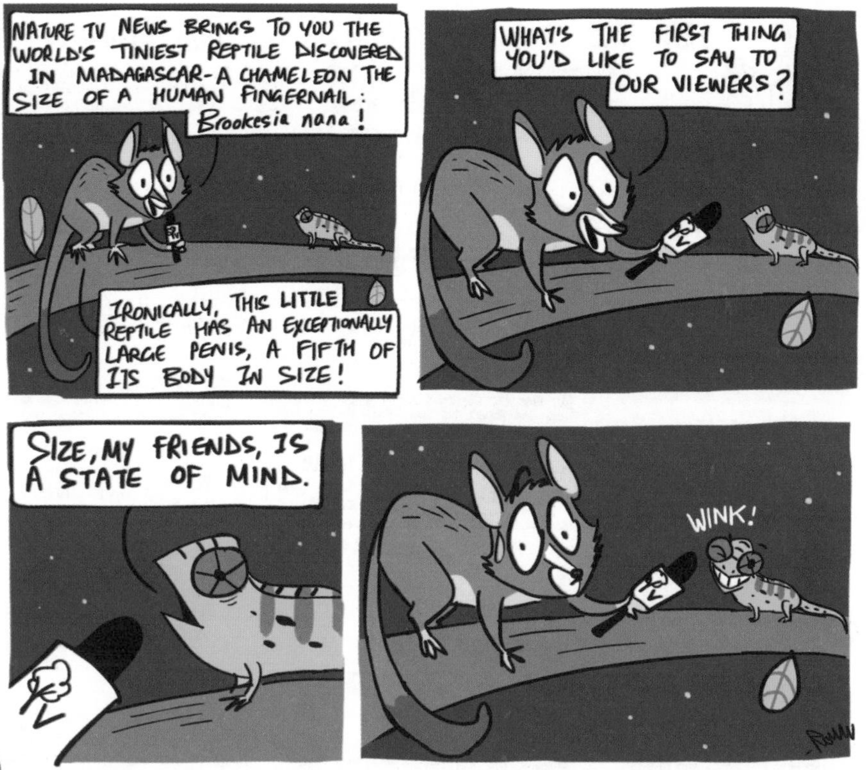
NATURE TV NEWS BRINGS TO YOU THE WORLD'S TINIEST REPTILE DISCOVERED IN MADAGASCAR- A CHAMELEON THE SIZE OF A HUMAN FINGERNAIL: Brookesia nana !
IRONICALLY, THIS LITTLE REPTILE HAS AN EXCEPTIONALLY LARGE PENIS, A FIFTH OF ITS BODY IN SIZE!
WHAT'S THE FIRST THING YOU'D LIKE TO SAY TO OUR VIEWERS?
SIZE, MY FRIENDS, IS A STATE OF MIND.
WINK!

#25

IT'S MONSOON IN THE WESTERN GHATS AND THE ENDEMIC SMALL TREE FROGS ARE MATING.
SOON AFTER THEY'VE DEPOSITED EGGS & SPERM OVER THE LEAF THEY MATED ON, THE MALE JUMPS AWAY.
TYPICAL.
THE FEMALE THEN BEGINS ROLLING THE LEAF INTO A PROTECTIVE, TUBE-LIKE NEST, STICKING IT AT THE ENDS WITH A VISCOUS FOAM.
HER HARD WORK ENSURES THAT THE EGGS DO NOT DESSICATE, REMAIN FIRM & MOIST, AND SAFE FROM PREDATORS.
PHEW. NOW I CAN ROLL ANOTHER ONE IN PEACE.

#26
HOW THE FIG WASP RUNS THE FIG EMPIRE
CEO
YOU MIGHT LOOK AT A FIG TREE AND THINK THAT IT HAS NO FLOWERS; BUT IT DOES. THEY'RE JUST HIDDEN FROM THE OUTSIDE WORLD.
SHOO! MY FLOWERS AREN'T TOO FOND OF MEDIA ATTENTION.
THE FIG 'FRUIT' IS ACTUALLY A FALSE FRUIT THAT ENCLOSES HUNDREDS OF THESE UNISEXUAL FLOWERS.
DO NOT TOUCH THE FLOWERS
WHAT FLOWERS?
THE CONCEALED FLOWERS ARE INACCESSIBLE TO MOST POLLINATORS. JUST ONE GROUP OF WASPS KNOWS HOW TO REACH THEM : THE FIG WASPS.
STEP ASIDE! THIS LOOKS LIKE A JOB FOR LADY FIGITA!
F
BUT FIGS ARE SELECTIVE ABOUT WHICH SPECIES OF WASP CAN ENTER THE FRUIT. EACH FIG SPECIES HAS ITS UNIQUE WASP POLLINATOR!
FIG-MENT, FIG-URINE, FIG-WORT- NONE OF THE COMBINATIONS WORKED.
HMMM...I SURE CAN'T SEEM TO FIG-URE THIS OUT.
ACCESS DENIED!
THE FEMALE WASP ENTERS THE FIG TO ST BUSINESS. ONCE INSIDE, SHE CANNOT CO OUT, AND DIES WITHIN THE FIG.
COULD YOU FETCH MY PURSE, PLEASE? THIS THING DOESN'T HAVE AN EXIT.
FIRST, THE FEMALE WASP POLLINATE A FEW OF THE FEMALE FLOWERS, WHICH MATURE BEFORE THE MALE FLOWERS. THIS INDUCES RIPENING OF THE FRUIT. THEN SHE LAYS HER EGGS ON THE OTHER FEMALE FLOWERS.
OKAY, GIRLS, WHO WANTS TO PLAY CATCH?
WASP LARVAE EMERGE FROM THE EGGS, FEEDING ON THE RIPENING FRUIT AND GROWING RAPIDLY.
MMM... FLAVOUR OF THE CENTURY!
ROASTED FIGS ON VANILLA
THE LARVAE METAMORPHOSE INTO ADULTS WITHIN THE FRUIT. ONLY THE ADULT FEMA HAVE WINGS. THE WINGLESS MALES WILL MATE AND DIE WITHIN THE FIG.
HEY! WANNA FLY OUT FOR SOME COFFEE?
OH. NEVE MIND.
AFTER MATING, THE FEMALE WASP COLLECTS POLLEN FROM THE MALE FLOWERS IN THE FIG, WHICH MATURE AROUND THE SAME TIME. THE MALE WASP CUTS OPEN A HOLE FOR THE FEMALE TO EXIT.
LISTEN, ROSE, YOU'RE GOING TO GET OUT OF HERE... YOU'RE GOING TO GO ON... & YOU'RE GOING TO MAKE LOTS OF BABIES... PROMISE ME.
THE FEMALE WASP FLIES OUT IN SEARCH OF ANOTHER FIG TREE TO POLLINATE ITS FEMALE FLOWERS WITH THE POLLEN, AND LAY HER EGGS, REPEATING THE CYCLE.
SO HOW'S YOUR DAY BEEN?
OH, HECTIC! I'VE BEEN LOADED!
WITH WORK?
NOPE. POLLEN.
THIS INTERDEPENDENCE OF THE FIG AN THE FIG WASP IS ONE OF NATURE'S MOST INTRICATE PARTNERSHIPS.
LOOSEN UP, DUDE. LITERALLY.

#27

FIREFLY COURTSHIP IN DARK HABITATS
...
...
FIREFLY COURTSHIP IN AREAS WITH LIGHT POLLUTION

#28

EYE FASHION BY INDIAN FROGS
THE SKY BLUE RIM BY THE BLUE-EYED BUSH FROG
'FLAME OF MY EYES' BY THE CHALAZODES BUSH FROG
'THE STRIPEY RETRO' BY THE STARRY-EYED TREE FROG
COMPLEMENTARY COLOUR RIM BY THE MALABAR GLIDING FROG
'SPOTLIGHTS' BY THE OCHLANDRA BUSH FROG
TRUST ME, LENSKART HASN'T EVEN HEARD OF OUR RANGE!

#29

THE DIFFERENT TYPES OF ANTS YOU'LL MEET IN AN ANT COLONY
THE QUEEN
HER JOB IS TO LAY EGGS AND TO KEEP THE COLONY GOING.
SHOBHA, WOULD YOU BELIEVE THIS?! I JUST LAID MY 300,001st EGG, BREAKING MY PREVIOUS RECORD!
POP!
THE LARVAE
WORKER ANTS FEED THE LARVAE WITH REGURGITATED CROP FLUID.
HEY! CAN'T YOU REGURGITATE PEPPERONI PIZZA FOR A CHANGE?
THE WORKERS
FEMALE ANTS THAT DO NOT REPRODUCE. THEIR JOB IS TO FEED THE COLONY AND CARE FOR THE LARVAE.
IT JUST SUCKS TO NEVER GET MATERNITY LEAVE!
I'M JUST WORRIED ABOUT MY PAY RAISE...
SHUT UP YOU TWO, AND RAISE THIS DARN THING!
THE SOLDIERS
WORKERS WITH LARGE HEADS & MANDIBLES. THEY FIGHT INTRUDERS AND PROTECT THE COLONY.
FOUND GUILTY OF REFUSING TO SAY, 'LONG LIVE THE QUEEN.'
TAKE HER TO THE INTERROGATION ROOM.
THE DIGGERS
WORKERS SPECIALIZING IN DIGGING AND EXPANSION OF THE COLONY.
I WAS SO WRONG WHEN I THOUGHT THAT LIFE WOULD BE EASIER AFTER MY PROMOTION!
THE DRONES
WINGED MALES THAT EMERGE WITH WINGED FEMALES FROM THE COLONIES, MATE WITH THEM & DIE. THEY ATTRACT FEMALES USING PHEROMONES.
I PROMISE TO LOVE YOU AS LONG AS I LIVE.
THAT'S SO ROMANTIC!
AND I'LL BE DEAD RIGHT AFTER WE MATE.
THE BREEDING FEMALES
WINGED FEMALES THAT MATE WITH MALES AND FLY IN SEARCH OF NEST SITES TO START THEIR OWN COLONIES.
ALL HAIL MARIE ANT-OINETTE THE NEW QUEEN OF FRANCE!
-ROHAN

#30

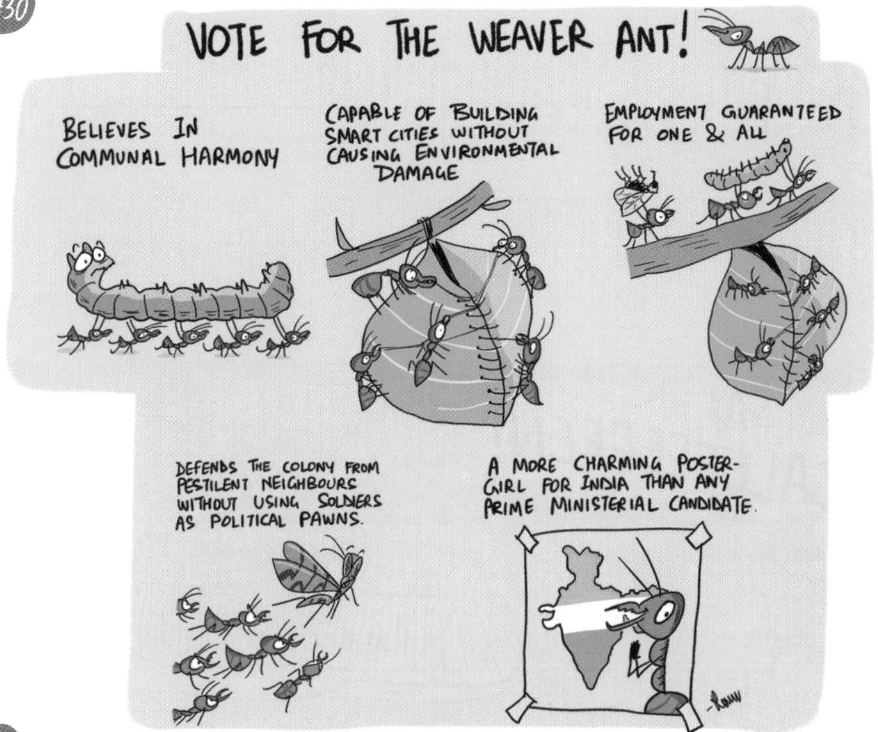
VOTE FOR THE WEAVER ANT!
BELIEVES IN COMMUNAL HARMONY
CAPABLE OF BUILDING SMART CITIES WITHOUT CAUSING ENVIRONMENTAL DAMAGE
EMPLOYMENT GUARANTEED FOR ONE & ALL
DEFENDS THE COLONY FROM PESTILENT NEIGHBOURS WITHOUT USING SOLDIERS AS POLITICAL PAWNS.
A MORE CHARMING POSTER-GIRL FOR INDIA THAN ANY PRIME MINISTERIAL CANDIDATE.

#31

AND HOW WOULD YOU LIKE YOUR EGGS, MA'AM— SCRAMBLED? POACHED? FRIED?
I'M THE INDIAN EGG-EATING SNAKE. SPECIAL ENAMEL-CAPPED PROJECTIONS ON MY CERVICAL VERTEBRAE EXTEND INTO MY OESOPHAGUS, HELPING ME CRACK EGGS AS I SWALLOW THEM...
LONG STORY SHORT, I'LL SCRAMBLE THEM MYSELF.
MM. OKAY.
ANYTHING TO... UMM... WASH IT DOWN WITH?
NO, THANKS. I'VE GOT ENZYMES.
CRUNCH
K.

SIX

Underwater

#1

SAY BALEEEEEEEN!
CLICK

As much as 71 per cent of the planet is covered by water. This means that 71 per cent of the planet is inhabited by wild animals that live a life we can only momentarily experience when we dive, snorkel or swim! It's truly another world underwater. From three-millimetre crabs to thirty-tonne whales and everything else in between, the whackiest of creatures make the blue planet blue. It's time to put on your scuba gear on and dive straight into a chapter that will make you 'Say Baleeeeen'.

#2

SOME WORLD RECORDS BY MARINE ANIMALS
LARGEST EXTANT INVERTEBRATE: COLOSSAL SQUID
AS IF MY NAME DIDN'T GIVE THAT AWAY ALREADY!
LONGEST GESTATION FOR A VERTEBRATE: FRILLED SHARK (3½ YEARS)
frills of pregnancy
Here's my 900th #mommytobe post!!!
FASTEST-SWIMMING SHARK AND HIGHEST-JUMPING FISH: SHORTFIN MAKO
WHEEEE!
MAKOS ARE WHACKOS!
LARGEST LIVING ANIMAL: BLUE WHALE
BEING A FILTER FEEDER ALSO MAKES ME THE WORLD'S LARGEST VACUUM CLEANER.
LOUDEST NATURAL SOUND: A PISTOL SHRIMP'S PINCER SNAP
YOU TALKIN' TO ME?
LARGEST BRAIN IN THE WORLD: SPERM WHALE
AND NATURALLY, THE MOST SAPIOSEXUAL OF ALL.
FASTEST FISH: SAILFISH
NO, I'M NOT THE CHEETAH OF THE SEA. THE CHEETAH IS THE SAILFISH OF LAND.
THE LARGEST FISH IN THE WORLD: WHALE SHARK
AND THE LARGEST CARPOOLING SERVICE!
LONGEST-LIVING MAMMAL: BOWHEAD WHALE (200+ YEARS)
I BOW MY HEAD TO MANKIND FOR COMPLETELY WRECKING THIS PLANET IN ONE BOWHEAD'S LIFETIME!
STRONGEST BIOLOGICAL MATERIAL: LIMPET TEETH
WHAT'S A DENTIST?
THE ONLY ORGANISM CAPABLE OF BIOLOGICAL IMMORTALITY: Turritopsis dohrnii (A JELLYFISH)
HOW HAS IT NEVER STRUCK MANKIND THAT THE GOD IT HAS BEEN SEEKING LIVES UNDER THE SEA?
THE WORLD'S FIRST KNOWN ANIMAL CAPABLE OF SURVIVING WITHOUT USING OXYGEN: Henneguya salminicola, A JELLYFISH-LIKE PARASITE OF SALMON.
NOT JUST GOD, THE DEVIL'S HERE TOO!
LONGEST PENIS IN PROPORTION TO BODY SIZE: THE BARNACLE
I MAY BE SESSILE, BUT GUESS WHAT ISN'T!
OOH BABY!
LONGEST-LIVING VERTEBRATE: GREENLAND SHARK (400 YEARS)
LOOK HOW FAST SCUBA HAS EVOLVED! FEELS LIKE I SHOOK HANDS WITH THAT COSTEAU CHAP JUST YESTERDAY!
LONGEST-LIVING ANIMAL: OCEAN QUAHOG CLAM (500+ YEARS)
HOW DOES ONE GO ON FOR THAT LONG?!
BY SHUTTING YOURSELF TO THE WORLD AND NEVER JOINING SOCIAL MEDIA.

#3

HAMMERHEAD HONEYMOON
WHACK!
BAMM!
SMACK!
WHAMM!
FORGET IT, SWEETHEART. WE CAN NEVER KISS.

#4

WAS CELEBRATING EARTH HOUR.

#5

OH MY GOD! DID YOU HEAR THAT?!
SHE SAID 'DADDA' IN SONAR!

#6

DEADLY CREATURES OF THE SEA
A SERIOUSLY SPINY LIONFISH!
A FATALLY VENOMOUS SEA SNAKE!
A PERILOUSLY POISONOUS FLATFISH!
AN ABOMINABLE STINGING JELLYFISH!
A LOATHFULLY LETHAL STINGRAY!
ALL OF THE ABOVE— ME, A SHAPE-SHIFTING MIMIC OCTOPUS!
OH, AND A VERY HAPPY APRIL FOOL'S TO YOU, AND TO ALL MY PREDATORS!

#7

WHAT YOU IMAGINE DOLPHIN SIGHTINGS TO BE LIKE-
WHAT WILD DOLPHIN SIGHTINGS REALLY ARE LIKE-

#8

THE DYNAMICS OF WHALE POOP

BEFORE WE TALK ABOUT WHALE POOP, REMEMBER THIS: WHALE FECES ARE REDDISH IN COLOUR DUE TO THE HIGH AMOUNT OF IRON IN THEM.
THE STOOL TEST REPORTS ARE PERFECTLY NORMAL, MA'AM. IN FACT, I MUST SAY, YOU'RE A WOMAN OF STEEL!
MD. PATH
THROUGH THEIR FECES, WHALES ACT AS ONE OF THE MAIN NUTRIENT RECYCLERS OF THE MARINE ECOSYSTEM, RESPONSIBLE FOR TRANSPORTING ESSENTIAL NUTRIENTS BETWEEN VARIOUS LEVELS OF THE SEA.
SIR, HERE'S THE LOGISTICS REPORT ON NITROGEN YOU'D ASKED FOR.
YOU'RE A DARLING, RITA.
MARINE NUTRIENT TRANSPORT CORP.
AREAS WITH WHALE POPULATIONS HAVE BEEN FOUND TO BE RICHER AND MORE PRODUCTIVE IN MARINE LIFE.
PARTY TIME, FOLKS! KRILL'S UP BY 85% THIS WEEK!
WOOHOO!
WHALES EAT A LOT OF KRILL. AND KRILL NEED IRON TO GROW. THIS IRON COMES FROM WHALE FECES!
THE LEPRECHAUN KRILL & HIS POT OF ~~GOLD~~ IRON ORE.
PHYTOPLANKTON, ON WHICH ALL MARINE LIFE DEPENDS, NEED IRON-RICH ENVIRONMENTS TO GROW, AND ARE HENCE ABUNDANT AROUND WHALE FECES. SMALL MARINE ANIMALS THAT FEED ON PHYTOPLANKTON THUS THRIVE AROUND WHALE FECES TOO, BOOSTING THE PRODUCTIVITY OF THE MARINE ECOSYSTEM.
THIS IS GOOD SHIT!
chomp
munch
chomp
ALTHOUGH WHALES FEED AT DEEPER LEVELS OF THE SEA, THEY STRATEGICALLY DEFECATE TOWARDS THE SURFACE, WHERE SUNLIGHT IS ABUNDANT, SO THAT PHYTOPLANKTON CAN USE THE SUNLIGHT & NUTRIENTS TO PROLIFERATE. THIS IS ALSO A MEANS OF TRANSPORTING NUTRIENTS FROM THE DEPTHS OF THE SEA TO THE SURFACE.
NOW YOU KNOW, HONEY, WHY MUMMY ALWAYS TEACHES YOU TO AIM HIGH?
KRILL FEED ON PHYTOPLANKTON AND MULTIPLY, THUS GENERATING FOOD FOR WHALES THEMSELVES. IN THIS WAY, WHALES HAVE ENGINEERED A SELF-SUFFICIENT ECONOMY.
BECAUSE WHEN YOU GENERATE YOUR OWN RESOURCES, THERE'S NOTHING LEFT TO FIGHT ABOUT.
NOBEL PRIZE for ECONOMICS
NOBEL PEACE PRIZE
AND THAT'S NOT ALL. PHYTOPLANKTON ALSO SEQUESTER CARBON, AND CHECK GLOBAL WARMING, INDIRECTLY MAKING WHALE POOP AN EFFECTIVE CARBON REGULATOR.
POLLUTION! HELP!!
HAVE NO FEAR! CARBON REGULA-TOR IS HERE!
YAY!
CR
FRRPPPP
CR
SO IN A NUTSHELL, WHALE POOP NOT ONLY HELPS MAINTAIN THE PHYTOPLANKTON BIOMASS OF THE OCEAN AND ENABLES WHALES TO 'FARM' THEIR OWN FOOD, BUT IT ALSO CURBS GLOBAL WARMING AND ENRICHES THE MARINE ENVIRONMENT.
LONG STORY SHORT, MY BUTT KNOWS MORE ECONOMICS THAN ALL YOUR UNIVERSITIES PUT TOGETHER.

#9

THE HISTORY OF THE LIGHT BULB
EDISON 1880
SWAN 1870
DAVY 1815
THE ANGLERFISH PALEOZOIC ERA

#10

THE FILM SOCIETY OF RAYS SCREENS ITS OLD FAVOURITE AUTEUR

Charulata
A FILM BY
SATYAJIT
RAY

#11

YOU KNOW, SON, ONE THING I'VE NEVER BEEN ABLE TO DEAL WITH MY ENTIRE LIFE, IS THE INFLATION. I HOPE YOU LEARN IT SOME DAY.

#12

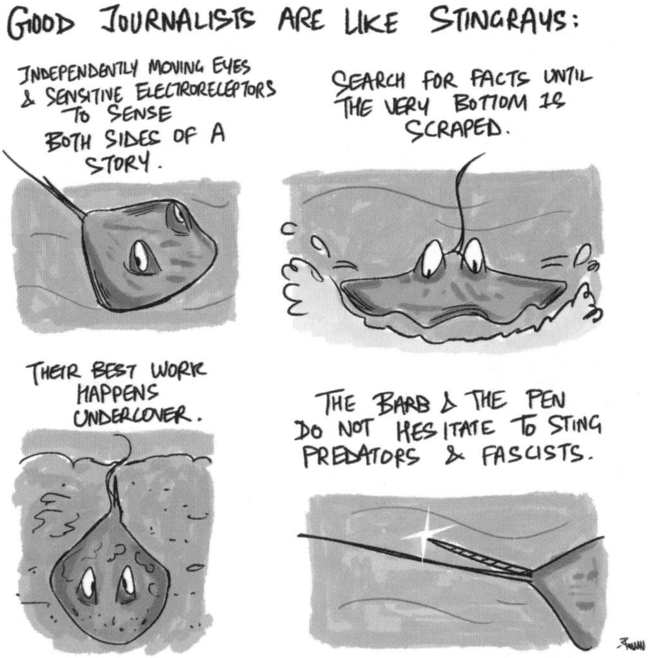
GOOD JOURNALISTS ARE LIKE STINGRAYS:
INDEPENDENTLY MOVING EYES & SENSITIVE ELECTRORECEPTORS TO SENSE BOTH SIDES OF A STORY.
SEARCH FOR FACTS UNTIL THE VERY BOTTOM IS SCRAPED.
THEIR BEST WORK HAPPENS UNDERCOVER.
THE BARB & THE PEN DO NOT HESITATE TO STING PREDATORS & FASCISTS.

#13

IF DECORATOR CRABS DID BRIDAL MAKE-UP
FIRST, A BACK DRAPE OF STINGING ANEMONE TO PROTECT YOU FROM UNWANTED ADVANCES...
NEXT, A PAIR OF INVERTED BACK CLAWS TO HELP YOU CARRY ALL YOUR ADORNMENTS...
NOW, SOME SPONGES TO CONCEAL YOUR LOVELY FOREARMS...
AND FINALLY, AN ALGAL CROWN FOR ADDED CAMOUFLAGE!
I'M BARELY NOTICEABLE.
ISN'T THAT THE WHOLE POINT, SWEETIE?

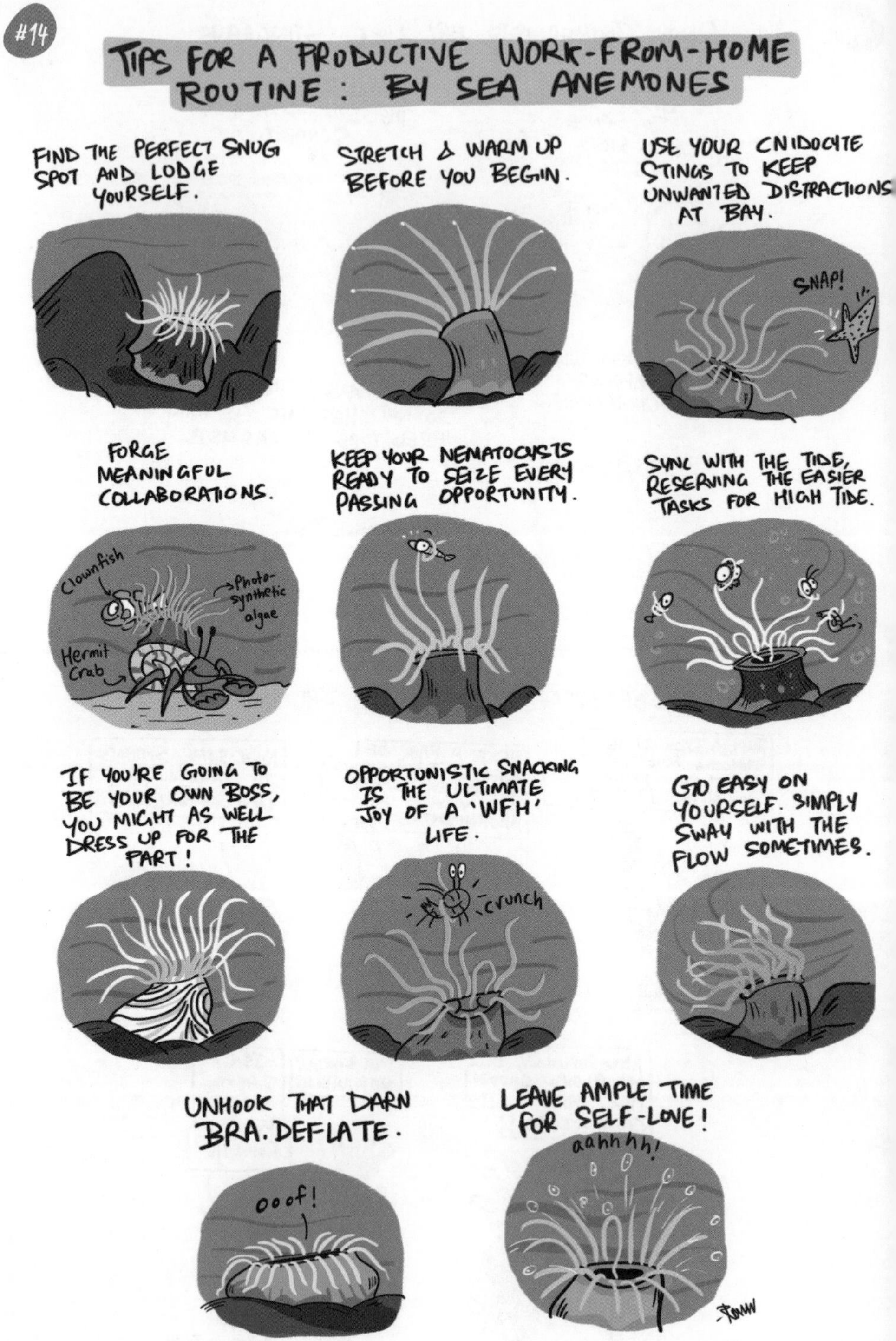
#14
TIPS FOR A PRODUCTIVE WORK-FROM-HOME ROUTINE: BY SEA ANEMONES
FIND THE PERFECT SNUG SPOT AND LODGE YOURSELF.
STRETCH & WARM UP BEFORE YOU BEGIN.
USE YOUR CNIDOCYTE STINGS TO KEEP UNWANTED DISTRACTIONS AT BAY.
SNAP!
FORGE MEANINGFUL COLLABORATIONS.
Clownfish
Photo-synthetic algae
Hermit Crab
KEEP YOUR NEMATOCYSTS READY TO SEIZE EVERY PASSING OPPORTUNITY.
SYNC WITH THE TIDE, RESERVING THE EASIER TASKS FOR HIGH TIDE.
IF YOU'RE GOING TO BE YOUR OWN BOSS, YOU MIGHT AS WELL DRESS UP FOR THE PART!
OPPORTUNISTIC SNACKING IS THE ULTIMATE JOY OF A 'WFH' LIFE.
crunch
GO EASY ON YOURSELF. SIMPLY SWAY WITH THE FLOW SOMETIMES.
UNHOOK THAT DARN BRA. DEFLATE.
ooof!
LEAVE AMPLE TIME FOR SELF-LOVE!
aahhhh!

#15

SECOND-HAND FASHION WITH THE HERMIT CRAB

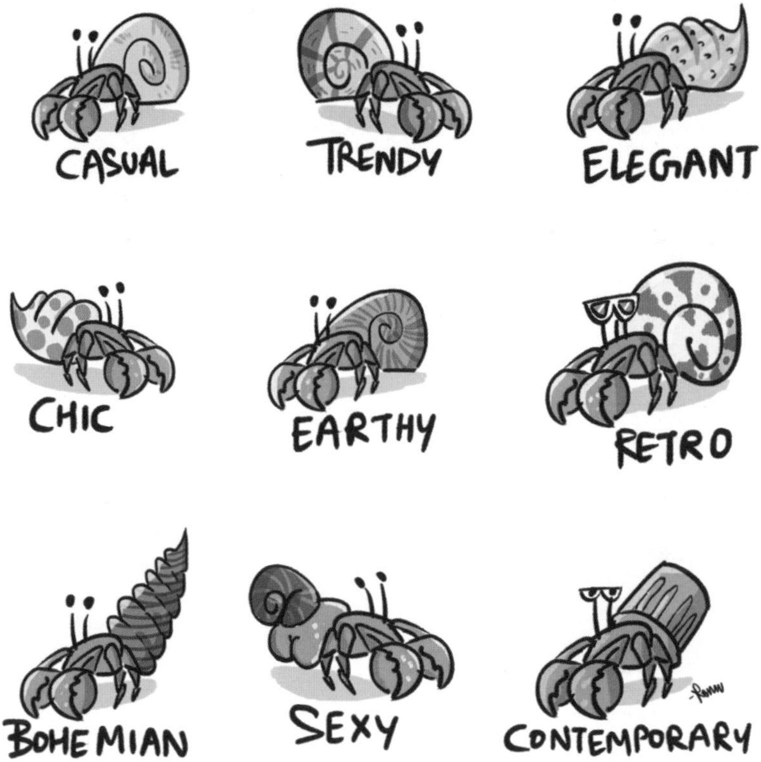
CASUAL
TRENDY
ELEGANT
CHIC
EARTHY
RETRO
BOHEMIAN
SEXY
CONTEMPORARY

#16

SO, ARE YOU A GOOGLE MAPS USER?
OR ARE YOU AN APPLE MAPS USER?
I'M A SEA TURTLE AND I'M AN EARTH'S MAGNETIC FIELD MAP USER.
THIS APP LETS ME DOWNLOAD ALL MY MAPS OFFLINE!

#17

SEA OTTERS HELP MAINTAIN KELP ECOSYSTEMS BY CONTROLLING THE NUMBER OF SEA URCHINS...
IN TURN, KELP IS USED BY SEA OTTER MOMS TO ANCHOR THEIR PUPS AND KEEP THEM AFLOAT, WHEN THEY LEAVE TO FORAGE!
THANKS FOR THE HELPING HAND, SEA OTTER!
THANKS FOR THE KELPING HAND, KELP!

#18

GOOD MORNING, CLASS! TODAY WE LEARN ABOUT A UNIQUE & SPECIAL FEATURE WE SHARKS ARE BLESSED WITH: THE AMPULLAE OF LORENZINI.
SHARK BIOLOGY
THESE AMPULLAE ARE A NETWORK OF ELECTRORECEPTORS IN OUR HEADS, WHICH HELP US DETECT ELECTRIC STIMULI FROM OUR PREY, MAKING HUNTING EASIER.
SLEEPER SHARK! YOU'RE DOZING IN MY CLASS AGAIN!
SLEEPER SHARK!
HUH? OH... UM... YES, HAMMERHEAD MA'AM.
PLEASE TELL THE CLASS WHAT 'AMPULLAE OF LORENZINI' ARE.
A KIND OF GOURMET PASTA.

#19

CORALS IN CARTOONS

I HAVE MY GROUPIES. I MEAN GROUPERS.
MARRY ME!
TAKE ME!

CORAL REEFS ARE NOT PLANTS! THEY'RE COMPLETE UNDERWATER ECOSYSTEMS FORMED AND HELD TOGETHER BY CALCIUM CARBONATE, SECRETED BY CORAL POLYPS.

I DON'T THINK I'VE SEEN A BETTER ART INSTALLATION IN THE WORLD!

CORALS OCCUPY LESS THAN 0.1% OF THE WORLD'S OCEAN SURFACE, AND ARE YET HOME TO MORE THAN A THIRD OF ALL MARINE SPECIES! THIS EARNS THEM THE TITLE 'RAINFORESTS OF THE SEA.'

OUR MOTTO: UNITY IN BIODIVERSITY.

INDIVIDUAL CORALS ARE CALLED POLYPS. THEY'RE AMONG THE SIMPLEST FORMS OF LIVING ORGANISMS, RELATED TO JELLYFISH & ANEMONE, MADE UP OF A SAC-LIKE BODY & TENTACLES.

YOUR FANS OFTEN WONDER HOW YOU MAKE ART WITHOUT HAVING EYES!
OH, IT JUST NEEDS VISION.

THERE ARE 3 MAIN TYPES OF CORAL REEFS-

FRINGING REEF: ATTACHED TO THE SHORE.

I PREFER THE SAFETY OF BABY POOLS.

BARRIER REEF: SEPARATED FROM THE SHORE BY A CHANNEL OR LAGOON.

DETACHMENT IS THE KEY TO ENLIGHTENMENT.
BOY, THE WEED MUST BE GREAT HERE!

ATOLL REEF: A RING-SHAPED REEF ENCIRCLING A LAGOON.

LOVED THE SEA, SO I PUT A RING ON IT.

CORALS COME IN AN ASTONISHING ARRAY OF SHAPES, SIZES & COLOURS!

PILLAR CORAL

YOUR COMPANY IS VERY REASSURING.

STAGHORN CORAL

WE HAVE BRANCHES WORLDWIDE.

FLUORESCENT CORAL

NO ONE TURNS OFF THE DISCO LIGHTS HERE!

BRAIN CORAL

NERD.

CORALS HAVE A SYMBIOTIC RELATIONSHIP WITH ZOOXANTHELLAE A KIND OF PROTOZOA, ZOOXANTHELLAE PROVIDE CORALS WITH NUTRIENTS & COLOUR (THROUGH PHOTOSYNTHESIS) AND IN TURN RECEIVE CARBON DIOXIDE & AMMONIUM FROM CORALS.

NEED A MAKEOVER? CONTACT ZOOXANTHELLAE, THE BEST BEAUTICIANS AROUND!

#20

BUT THIS ASSOCIATION IS VERY SUSCEPTIBLE TO RISING SEA TEMPERATURES. GLOBAL WARMING CAUSES ZOOXANTHELLAE TO GET EXPELLED FROM CORALS, RESULTING IN A LOSS OF COLOUR & NUTRIENTS, AND EVENTUALLY THE DEATH OF THE REEF. THIS PHENOMENON IS CALLED CORAL BLEACHING.
THIS PLACE LOOKED LIKE MOULIN ROUGE ONCE. NOW IT'S A HORROR FILM SET.
CORAL REEFS ARE NOT ONLY A HAVEN FOR MARINE LIFE, BUT THEY ARE ALSO INDISPENSABLE TO HUMANS. THEY KEEP SHORELINES INTACT, AND CONTRIBUTE BILLIONS OF DOLLARS ANNUALLY TO THE GLOBAL ECONOMY THROUGH TOURISM AND FISHERIES.
@#$%.! THE REEF IS BLEACHED!
SO IS THE WORLD BANK.
REEFS WORLDWIDE ARE UNDER THREAT FROM CLIMATE CHANGE, OCEAN ACIDIFICATION, BLAST-FISHING, POLLUTION AND OVER-HARVESTING OF REEF RESOURCES. THE GREAT BARRIER REEF, THE WORLD'S LARGEST CORAL REEF, SUFFERED A RECENT MASS BLEACHING EVENT.
WHAT ELSE IS FOUND IN THE SEA AND IS 'GREAT WHITE'?
WHAT?
THE GREAT BARRIER REEF.

#21

WHY DON'T YOU WAIT HERE WHILE I GO CHANGE INTO SOMETHING MORE 'COMFORTABLE'?
OOH! SURE!
♩
WHOA MAMA! WASN'T EXPECTING THAT!
SORRY, MY SHELL ERODED COMPLETELY FROM OCEAN ACIDIFICATION.

#22

MEET INDIA'S MOST SUSTAINABLE DRESS DESIGNER – THE DECORATOR WORM!
THIS MARINE WORM, A COMMON SIGHT ON OUR SHORES, RESIDES IN A POLYSACCHARIDE TUBE.
THE WORM COLLECTS EMPTY SHELLS, PEBBLES & EVEN TRASH, BINDING THEM AROUND THE TUBE INTO A DAZZLING PROTECTIVE COSTUME!
WOW! WILL YOU PLEASE DESIGN COSTUMES FOR MY NEXT FILM?
SURE, MR. BHANSALI. PERIOD OR CONTEMPORARY?

#23

KILLER WHALE! WHY WOULD ANYONE CALL YOU THAT? YOU'RE NOTHING LIKE THAT NAME!
FRIEND, THANKS FOR UNDERSTANDING! HOW COULD ANYONE SEE A KILLER IN SOMETHING AS GENTLE AS ME?!
NO, I'M TALKING ABOUT THE 'WHALE' BIT. GENETICALLY, YOU'RE A DOLPHIN.
@#*%ING BIOLOGY NERDS.

#24

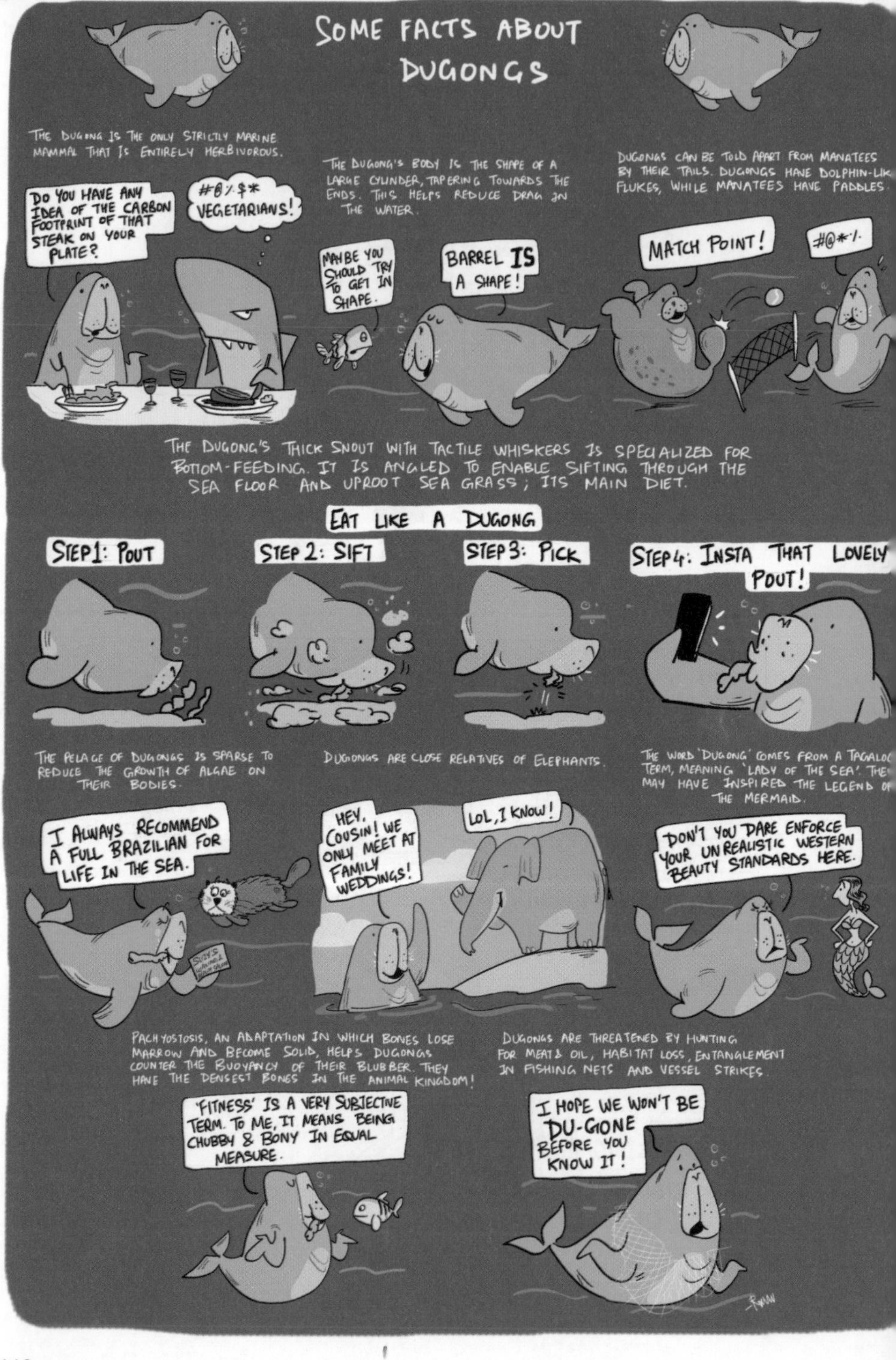

SOME FACTS ABOUT DUGONGS
THE DUGONG IS THE ONLY STRICTLY MARINE MAMMAL THAT IS ENTIRELY HERBIVOROUS.
DO YOU HAVE ANY IDEA OF THE CARBON FOOTPRINT OF THAT STEAK ON YOUR PLATE?
#@%$* VEGETARIANS!
THE DUGONG'S BODY IS THE SHAPE OF A LARGE CYLINDER, TAPERING TOWARDS THE ENDS. THIS HELPS REDUCE DRAG IN THE WATER.
MAYBE YOU SHOULD TRY TO GET IN SHAPE.
BARREL IS A SHAPE!
DUGONGS CAN BE TOLD APART FROM MANATEES BY THEIR TAILS. DUGONGS HAVE DOLPHIN-LIK FLUKES, WHILE MANATEES HAVE PADDLES.
MATCH POINT!
#@*%.
THE DUGONG'S THICK SNOUT WITH TACTILE WHISKERS IS SPECIALIZED FOR BOTTOM-FEEDING. IT IS ANGLED TO ENABLE SIFTING THROUGH THE SEA FLOOR AND UPROOT SEA GRASS; ITS MAIN DIET.
EAT LIKE A DUGONG
STEP 1: POUT
STEP 2: SIFT
STEP 3: PICK
STEP 4: INSTA THAT LOVELY POUT!
THE PELAGE OF DUGONGS IS SPARSE TO REDUCE THE GROWTH OF ALGAE ON THEIR BODIES.
I ALWAYS RECOMMEND A FULL BRAZILIAN FOR LIFE IN THE SEA.
DUGONGS ARE CLOSE RELATIVES OF ELEPHANTS.
HEY, COUSIN! WE ONLY MEET AT FAMILY WEDDINGS!
LOL, I KNOW!
THE WORD 'DUGONG' COMES FROM A TAGALO TERM, MEANING 'LADY OF THE SEA'. THE MAY HAVE INSPIRED THE LEGEND O THE MERMAID.
DON'T YOU DARE ENFORCE YOUR UNREALISTIC WESTERN BEAUTY STANDARDS HERE.
PACHYOSTOSIS, AN ADAPTATION IN WHICH BONES LOSE MARROW AND BECOME SOLID, HELPS DUGONGS COUNTER THE BUOYANCY OF THEIR BLUBBER. THEY HAVE THE DENSEST BONES IN THE ANIMAL KINGDOM!
'FITNESS' IS A VERY SUBJECTIVE TERM. TO ME, IT MEANS BEING CHUBBY & BONY IN EQUAL MEASURE.
DUGONGS ARE THREATENED BY HUNTING FOR MEAT & OIL, HABITAT LOSS, ENTANGLEMENT IN FISHING NETS AND VESSEL STRIKES.
I HOPE WE WON'T BE DU-GONE BEFORE YOU KNOW IT!

#25
SOME PLEASURES OF TIDE-POOLING
HIP-SWAYING ANEMONES
SAND BUBBLER CRAB ART EXHIBITIONS
CANNES-READY HERMIT CRABS
AWKWARDLY SMILING MORAYS
A DIFFERENT KIND OF STAR-GAZING
INTERIOR DECOR BY ZOANTHIDS AND HYDROIDS
ECO-FRIENDLY HOLI WITH BIVALVES
BEACH PARTIES THAT EVEN INTROVERTS CAN ENJOY
EVERY BEACH IS A NUDIST BEACH WITH NUDIBRANCHS AROUND!

#26

USES OF THE WHALE SHARK'S POLKA DOTS
FOR CAMOUFLAGE, BY MIMICKING SCATTERED LIGHT UNDERWATER
CONVENIENT CASUAL-FRIDAY DRESSING
MAKING A BOLD FASHION STATEMENT
PLAYING 'CONNECT-THE-DOTS' WHEN BORED
THE WORLD'S LARGEST CHINESE CHECKERBOARD
RWW

#27

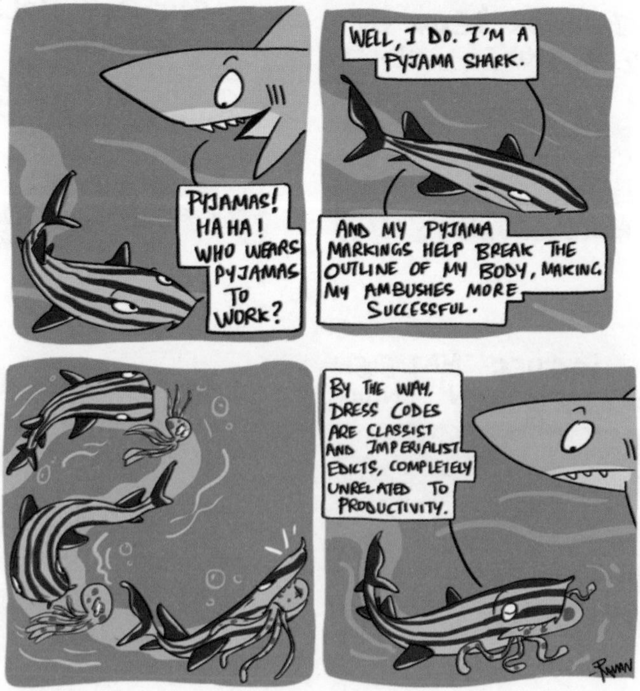
PYJAMAS! HAHA! WHO WEARS PYJAMAS TO WORK?
WELL, I DO. I'M A PYJAMA SHARK.
AND MY PYJAMA MARKINGS HELP BREAK THE OUTLINE OF MY BODY, MAKING MY AMBUSHES MORE SUCCESSFUL.
BY THE WAY, DRESS CODES ARE CLASSIST AND IMPERIALIST EDICTS, COMPLETELY UNRELATED TO PRODUCTIVITY.
RWW

#28

tap tap
Ucabs
Cheap underwater rides
Book Now
Pick-up location:
Brain Coral Cluster #8
Drop location:
Staghorn Apartments
AVAILABLE RIDES:
PRIVATE SEDAN
PRIVATE MINI
SHARED MINI
Pay using:
Cash
Debit Card
Mobile Wallet
Volunteer to clean Cab
Your ride has arrived!
MADAM, O.T.P?
UHH, ONE MOMENT PLEASE...

#29

Ang-Ler
BIOLUMINESCENCE IS #litAF!

#30

ONE TURTLE, MANY RECORDS: THE LEATHERBACK TURTLE

THE LEATHERBACK TURTLE IS THE LARGEST LIVING TURTLE, AND THE FOURTH-LARGEST LIVING REPTILE IN THE WORLD!

MARLIN STUDIOS PRESENTS
LEATHERBACK TURTLE
IN & AS
THE INCREDIBLE
BULK 3D!

LEATHERBACK TURTLES HAVE THE LARGEST FLIPPERS FOR ANY TURTLE.

NEXT!
UNDERWATER ARM-WRESTLING CHAMPIONSHIP

THE LEATHERBACK DIFFERS FROM OTHER TURTLES IN LACKING A BONY SHELL. INSTEAD OF SCUTES, IT HAS THICK, LEATHERY SKIN.

LOVE HANDLES, THE CARTOONIST MEANS TO SAY.

LEATHERBACKS ARE UNIQUE AMONG REPTILES IN THEIR ABILITY TO MAINTAIN HIGH BODY TEMPERATURES USING METABOLICALLY GENERATED HEAT.

OOOH! SOMEBODY'S SIZZLING TONIGHT!
IT'S MY METABOLIC RATE.

LEATHERBACK TURTLES ARE AMONG THE DEEPEST DIVING MARINE ANIMALS, DIVING AS DEEP AS 4200 FEET!

TCH TCH. IS THAT THE BEST YOU CAN DO?
20,000 LEAGUES UNDER THE SEA
-Jules Verne

LEATHERBACK TURTLES ARE ALSO THE FASTEST MOVING REPTILES, SWIMMING AT 35.28 KMPH!

#fatisthe newfast

LEATHERBACKS HAVE THE WIDEST DISTRIBUTION OF ALL SEA TURTLES, OCCURRING THROUGHOUT THE WORLD IN SUITABLE HABITATS, EXTENDING EVEN INTO THE ARCTIC CIRCLE.

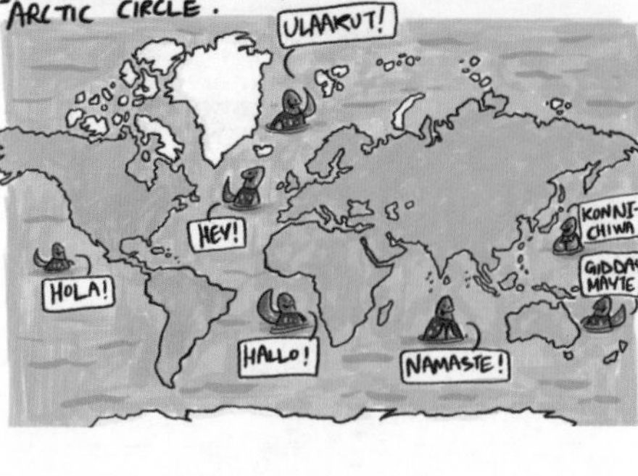

ULAAKUT!
HEY!
HOLA!
KONNICHIWA!
GIDDAY, MAYTE!
HALLO!
NAMASTE!

LEATHERBACKS FEED ALMOST EXCLUSIVELY ON JELLYFISH AND PLAY AN IMPORTANT ROLE IN CONTROLLING THEIR NUMBERS. BUT THIS ALSO MAKES THEM EXTREMELY SUSCEPTIBLE TO CONSUMING PLASTIC.

ATTENTION, KIDS. YOUR LIVES DEPEND ON TODAY'S LESSON!
BIOMIMICRY
JELLYFISH ✓
PLASTIC BAG ✗

SOME FACTS ABOUT PARROTFISH
PARROTFISH ARE NAMED AFTER THEIR UNIQUE DENTITION, A SET OF STRONG TEETH PACKED TIGHTLY, MAKING THE JAW APPEAR LIKE A PARROT'S BEAK.
HEY! WEREN'T YOU IN THAT TOOTHPASTE AD?
I'M IN EVERY UNDERWATER TOOTHPASTE AD.
PARROTFISH USE THEIR STRONG JAWS & TEETH TO PULVERIZE CORALS AND FEED ON ALGAE.
IN OTHER WORDS, I SPECIALIZE IN GIVING HICKIES TO CORALS.
YEAOW!
CHOMP!
BUT THIS DOESN'T MEAN THAT THEY'RE HARMFUL FOR CORALS. IN FACT, PARROTFISH ARE REEF GARDENERS, FEEDING ON ALGAE AND DEAD CORAL, AND KEEPING THE REEF IN GOOD HEALTH.
SHE SURE KNOWS ALL MY SPOTS!
CHOMP!
PARROTFISH EXCREMENT FROM ALL THE CORAL-MUNCHING IS WHAT FORMS MOST OF THE SAND IN TROPICAL BEACHES WORLDWIDE!
#BEACHLIFE
ACTUALLY IT'S HASHTAG PARROTFISH-POOP-LIFE.
AMONG THE 95 SPECIES OF PARROTFISH, THE GREEN HUMPHEAD PARROTFISH IS THE LARGEST AND THE BLUELIP PARROTFISH THE SMALLEST.
NICE LIP-GLOSS!
NICE...UMM... HELMET.
PARROTFISH HAVE THE MOST GAUDY FASHION SENSE, SPORTING BRIGHT, IRIDESCENT & BIZARRE COLOURS.
CANNES & MET GALA HAVE SEEN NOTHING, DARLING.
PARROTFISH HAVE COMPLICATED SEX LIVES, WITH MANY SPECIES BEING SEQUENTIAL HERMAPHRODITES, STARTING LIFE AS FEMALES AND LATER CHANGING SEX TO MALE.
BEING COLOURFUL ACTUALLY COMES FROM WITHIN.
THE CONSERVATION OF PARROTFISH IS CRUCIAL FOR SUSTAINING HEALTHY REEFS, AND THEREFORE, HEALTHY FISHERIES, OCEANS, AND OF COURSE, BEAUTIFUL BEACHES!
MEMORIZE THAT. LIKE A PARROT.
ROHAN

#32

THE DORSAL FIN OF THE REMORA IS MODIFIED INTO A SUCTION DISC.
THIS ENABLES THE REMORA TO ATTACH ITSELF TO LARGE HOSTS, SUCH AS SHARKS. THE REMORA GETS A FREE RIDE & PROTECTION, AND IN TURN CLEANS THE HOST'S BODY OF PARASITES.
ALL THAT'S GREAT, REMORA...
BUT MAYBE YOU SHOULD STOP LATCHING ON TO YOUR EX LIKE THAT.
Kiss
NOPE.

#33

I'M THE ENDANGERED NAPOLEON WRASSE, AND HERE'S A LIST OF CONSERVATION THREATS I FACE—
INTENSIVE HARVESTING FOR THE LIVE FISH TRADE ACROSS SOUTH EAST ASIA
BLAST FISHING AND CYANIDE FISHING
OCEAN ACIDIFICATION AND CORAL BLEACHING
CAPTURE FOR MARINE AQUARIUMS
OVERFISHING
WELL, I GUESS THIS IS MY WATERLOO.

#34

SO, GREENLAND SHARKS HAVE THE LONGEST LIFESPAN FOR ANY VERTEBRATE – 400 YEARS...
AND THE SPECIES WAS DESCRIBED TO SCIENCE BY MARCUS BLOCH AND J.G. SCHNEIDER IN 1801.
WHICH MEANS THAT GREENLAND SHARKS FROM THE BATCH THAT WAS FIRST DESCRIBED ARE STILL ROAMING THE NORTH-ATLANTIC!
ever_green
Major throwback to 1801– me & besties Bloch & Schneider celebrating a historic moment in icthyology. miss you guys!
#vintage #nofilter #tbt

#35

A LEAF IN THE MIDDLE OF THE SEA?!
ACTUALLY, I'M A GREEN SEA SLUG. I DON'T JUST LOOK LIKE A LEAF; I LIVE LIKE ONE TOO– I PHOTOSYNTHESIZE!
BUT AREN'T YOU EATING ALGAE?
chomp chomp
THAT'S RIGHT. IN A PROCESS CALLED KLEPTOPLASTY, I STEAL CHLOROPLASTS FROM THE ALGAE I EAT. THESE CHLORO-PLASTS GIVE ME ENERGY FROM PHOTOSYNTHESIS WHEN ALGAE ARE IN SHORT SUPPLY.
AND WHAT'S MORE, THE GREEN COLOUR THAT THE CHLOROPLASTS LEND MY BODY AIDS CAMOUFLAGE.
AND OF COURSE, THERE'S NO GREATER JOY THAN TO HAVE A DIET WITH A ZERO CARBON FOOTPRINT.
WOW. AND I THOUGHT VEGANISM IS EXTREME!

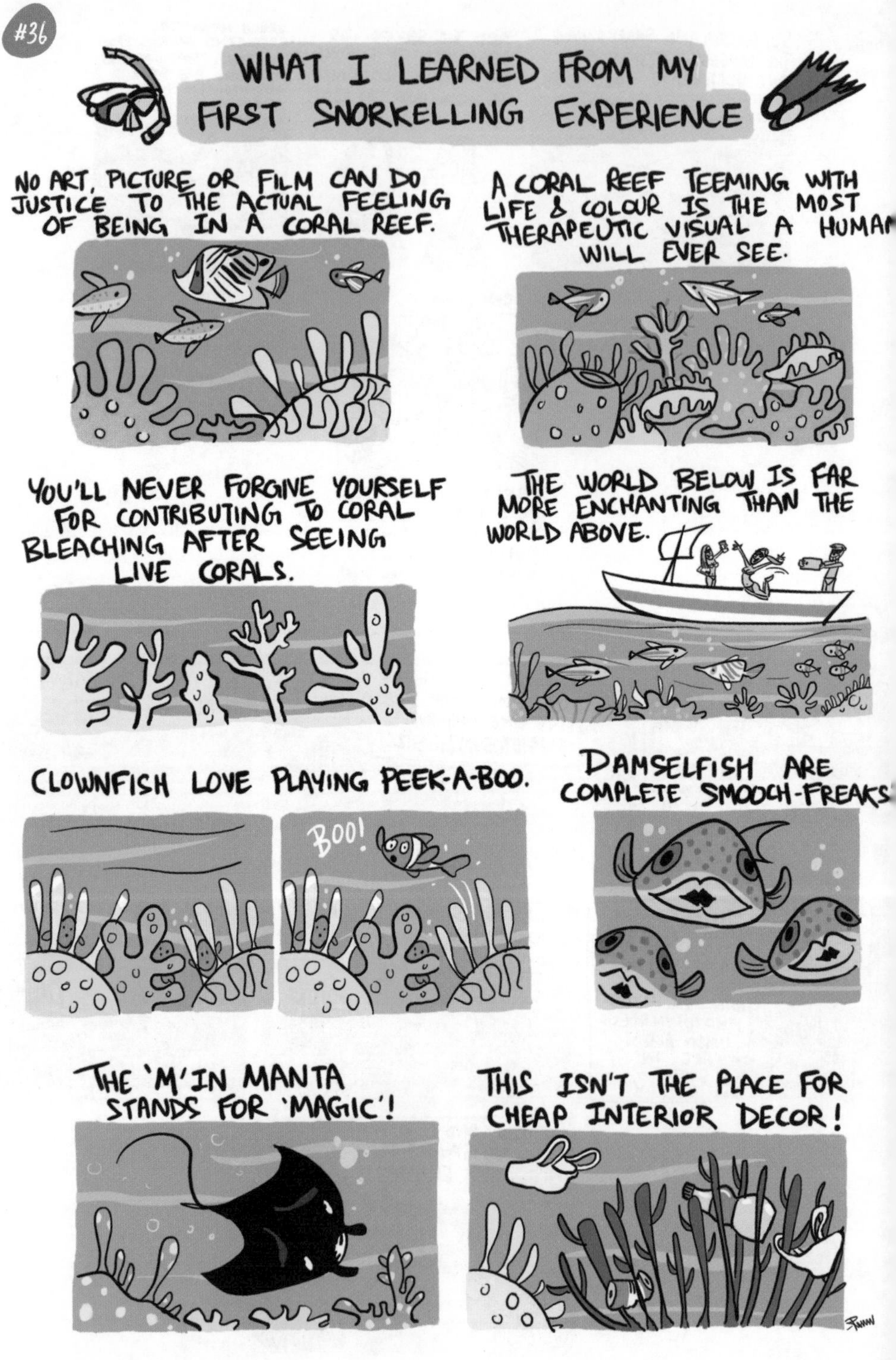
#36
WHAT I LEARNED FROM MY FIRST SNORKELLING EXPERIENCE
NO ART, PICTURE OR FILM CAN DO JUSTICE TO THE ACTUAL FEELING OF BEING IN A CORAL REEF.
A CORAL REEF TEEMING WITH LIFE & COLOUR IS THE MOST THERAPEUTIC VISUAL A HUMAN WILL EVER SEE.
YOU'LL NEVER FORGIVE YOURSELF FOR CONTRIBUTING TO CORAL BLEACHING AFTER SEEING LIVE CORALS.
THE WORLD BELOW IS FAR MORE ENCHANTING THAN THE WORLD ABOVE.
CLOWNFISH LOVE PLAYING PEEK-A-BOO.
BOO!
DAMSELFISH ARE COMPLETE SMOOCH-FREAKS!
THE 'M' IN MANTA STANDS FOR 'MAGIC'!
THIS ISN'T THE PLACE FOR CHEAP INTERIOR DECOR!

#37

MEET THE COOKIECUTTER SHARK
HEY, WHALE! IN THE MOOD FOR SOME COOKIES?
WOW, SURE!
CHOMP!
ONE FOR ME & ONE FOR YOU!
YEEAAOWW!

ON NATURE TV NEWS TODAY, WE CONGRATULATE THE COMMON CUTTLEFISH FOR BEING THE FIRST KNOWN INVERTEBRATE TO POSSESS THE VIRTUE OF SELF-CONTROL!
PASSING THE MARSHMALLOW TEST & CHOOSING SELF-RESTRAINT OVER INSTANT GRATIFICATION WAS ONCE CONSIDERED THE CORNERSTONE OF VERTEBRATE INTELLIGENCE. HOW DO YOU FEEL BREAKING THIS GLASS CIELING?
VERTEBRATE INTELLIGENCE? REALLY?
AS WE SPEAK, AN ENTIRE GENERATION OF INTELLIGENT BIPED VERTEBRATES IS CHASING INSTANT GRATIFICATION ON INSTAGRAM.

SEVEN
Flora

FOREST FIRE!!
WHAT! WHERE?!
JUST KIDDING!
#1
THE FLAME OF THE FOREST IS IN FULL BLOOM!

My job as a cartoonist who draws about nature is actually two profiles rolled into one. One is being a cartoonist and the other a naturalist. One of India's finest ecologists once said to me, 'One can never be a good naturalist without knowing trees in and out.' And the fact that I look at trees only to look at the animal or bird perched atop them makes me a rather terrible naturalist. Don't get me wrong; I have nothing against trees. It's just that the way botany presents itself to laymen doesn't quite amuse me like birds do. And trust me, I'm working on it. Until then, I hope you'll find my cartoons about the few trees and plants I've chosen to draw a little more amusing than how I find botany.

#2

FREE FOOD...
FREE SHELTER...
MASS EMPLOYMENT...
SECURITY...
EDUCATION...
A BOOMING ECONOMY
SURPLUS PROFIT...
AND LUXURY COMMUNITY HOUSING...
I'D VOTE FOR THE FIG TREE IF IT RAN FO PRIME MINISTE

SOME COMMON NEIGHBOURHOOD TREES OF DELHI AND THEIR SPECIALITIES

RED SILK COTTON

A FAVOURITE AMONG NECTAR-LOVING WILDLIFE, MY COTTON-LIKE CAPSULES CAN BE SEEN FLOATING IN THE AIR IN SUMMERS!

AMALTAS

MY PROFUSE YELLOW FLOWERING EARNS ME THE TITLE 'GOLDEN SHOWER TREE'.

ARAK

MY BARK IS USED AS A NATURAL TOOTHBRUSH, 'MESWAK'!

SHAMI

WORSHIPPED DURING DUSSEHRA, I WAS THE FOUNDATION FOR THE CHIPKO MOVEMENT.

PALASH

MY BRIGHT-RED, FIRE-SHAPED FLOWERS EARN ME THE TITLE, 'FLAME OF THE FOREST'!

SAUSAGE TREE

WHEN DELHI CHOKES AT NIGHT, MY RED FLOWERS BLOOM AND REFRESH THE AIR, ATTRACTING BATS THAT POLLINATE THEM.

CHEESEWOOD TREE

MY FLOWERS BLOOM IN THE WINTER AND PERFUME DELHI'S STREETS WHEN AIR POLLUTION IS AT ITS PEAK.

ANOTHER BONUS FUN FACT: **WE** ARE THE AIR PURIFIERS DELHI SHOULD BE INSTALLING, INSTEAD OF THIS JUNK.

#4

OKAY, HERE WE GO- ONE... TWO... AND
TADAAA!
OH.MY.GOD!
WHERE ARE WE?!
THE KUEKENHOF GARDENS IN THE NETHERLANDS?!
THE LAVENDER FIELDS OF PROVENCE?!
THE VALLEY OF FLOWERS?!
NOPE! THESE ARE THE 'KARVI' FLOWERS THAT BLOOM ONCE IN EIGHT YEARS...
AND WE... ARE IN MUMBAI!
NO WAY! ARE WE DRUNK?!
NOT... YET!
SANJAY GANDHI NATIONAL PARK
GLUG
GLUG

#5
NEELAKURINJI FLOWERS THAT BLOOM ONCE IN 12 YEARS ARE NOW BLOOMING IN MUNNAR, KERALA!
THE BLOOM HAS TRANSFORMED THE HILLS INTO SCENIC CARPETS OF LUSH LAVENDER!
AND HAS SENT ENDEMIC WILDLIFE INTO A SELFIE-FRENZY! HERE ARE SOME UPDATES FROM KERALA'S VERY OWN INSTA INFLUENCERS:
THE NILGIRI PIPIT
TOTALLY GOING YIPPITEE-YIPPIT!
THE CRIMSON-BACKED SUNBIRD
#YOLO IN 12 YEARS!
THE NILGIRI LANGUR
#accessorynotsorry
THE NILGIRI TAHR
#couplegoals
THE SOUTHERN BIRDWING BUTTERFLY
#happyhours!
THE INDIAN ELEPHANT
Where else but #godsowncountry!
-ROHAN

#6

HELLO, NEIGHBOUR! I'M YOUR FRIENDLY NEIGHBOURHOOD TREE.
I OFFER YOU SHADE WHEN YOU SIP YOUR COFFEE...
I SERVE AS THE STAGE FOR THE PRINIA'S PERFORMANCE THAT ENTHRALLS YOU EVERY MORNING...
I'M HOME TO THE PALM SQUIRRELS WHOSE ANTICS YOU LOVE TO WATCH...
I PROVIDE FREE LESSONS IN ENTOMOLOGY WHEN YOU'RE IN ONE OF YOUR CURIOUS MOODS...
I GIVE YOU COVER FROM NOSY NEIGHBOURS WHEN YOU'RE HAVING A PRIVATE MOMENT IN THE BALCONY...
OR WHEN YOU'RE SCRATCHING AN AWKWARD ITCH...
I'M ALSO YOUR PETS' FAVOURITE TOILET SPOT!
BUT DO YOU KNOW WHAT SPECIES I BELONG TO? WHAT COLOUR MY FLOWERS ARE? WHICH MONTH I BEAR FRUITS IN?
WELL... FOR AN INDIAN NEIGHBOUR YOU DON'T SEEM VERY INQUISITIVE.

#7

THE MANGROVE JANTA PARTY (MA.JA.PA)
DEAR CITIZENS OF MUMBAI, A LOVING NAMASKAAR.
CAN YOU NAME THE PARTY THAT HAS SERVED MUMBAI FOR THE LONGEST TERM?
I'LL TELL YOU. THE MA.JA.PA.!
WHICH PARTY HAS PROTECTED THIS CITY'S COASTLINE FROM FLOOD & EROSION FOR AS LONG AS IT HAS EXISTED?
THE MA.JA.PA.!
WHICH PARTY HAS SILENTLY BEEN ABSORBING THIS CITY'S CARBON EMISSIONS, WITHOUT EVEN LEVYING TAXES?
THE MA.JA.PA.!
AND WHICH PARTY HAS BEEN PRESERVING COASTAL BIODIVERSITY & CREATING LIVELIHOODS FOR COASTAL COMMUNITIES FOR CENTURIES?
YES, THE MA.JA.PA.!
AND NOW, THE GOVERNMENT WANTS TO UPROOT MUMBAI'S LONGEST-SERVING PARTY, TO MAKE WAY FOR WHAT IT CALLS 'DEVELOPMENT'! CAN MUMBAI LET THIS HAPPEN?
THESE AERIAL ROOTS HAVE ALWAYS ROOTED FOR MUMBAI. NOW IT'S MUMBAI'S TURN TO ROOT FOR US!
SO PLEASE. VOTE FOR THE MA.JA.PA.
OR ELSE, GOOD LUCK FOR THE TSUNAMI NEXT YEAR.

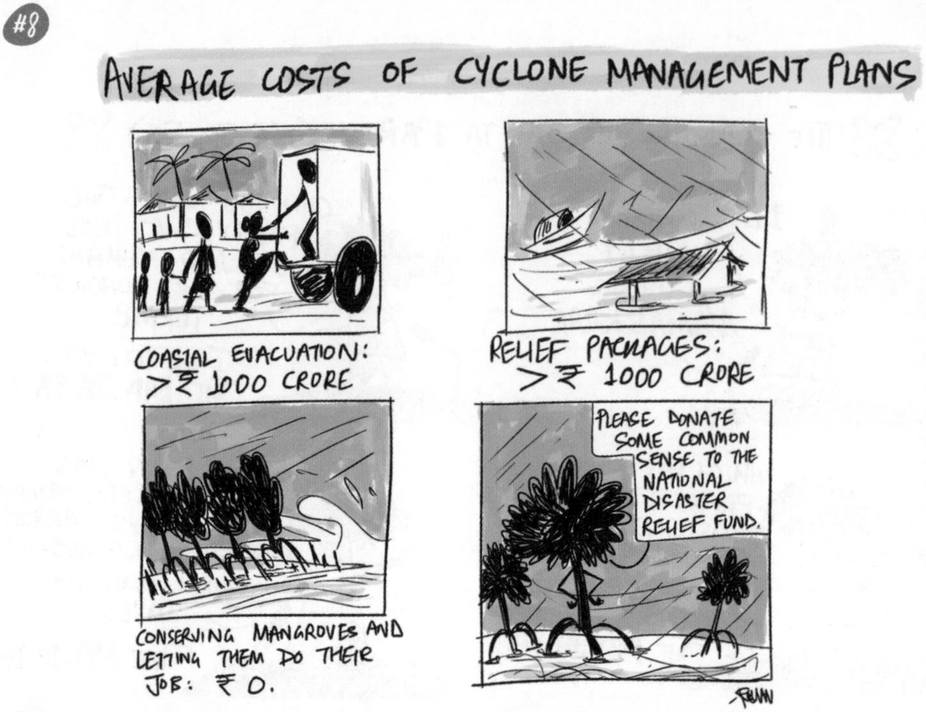
#8
AVERAGE COSTS OF CYCLONE MANAGEMENT PLANS
COASTAL EVACUATION: >₹ 1000 CRORE
RELIEF PACKAGES: >₹ 1000 CRORE
CONSERVING MANGROVES AND LETTING THEM DO THEIR JOB: ₹ 0.
PLEASE DONATE SOME COMMON SENSE TO THE NATIONAL DISASTER RELIEF FUND.
RUHN

#9
WHITE CHIPPI, HOW DO YOU FEEL ON BEING NAMED MAHARASHTRA'S STATE MANGROVE TREE ?
I'LL CELEBRATE ONLY WHEN THE GOVERNMENT STOPS DESTROYING MANGROVES & COASTS FOR LINEAR INTRUSIONS.
I CAN ONLY TOLERATE SALINITY, NOT DOUBLE STANDARDS.
RUHN

#10

OIL PALM MONOCULTURE
VS
RAINFORESTS OF NORTH-EAST INDIA
I'VE CONQUERED SO MANY TROPICAL COUNTRIES! NOW, IT'S INDIA'S TURN!
OH NO, YOU DON'T. IN A LAND OF SUCH DIVERSE CULTURES, THERE'S NO PLACE FOR MONOCULTURE. HA!
TOUCHÉ! BUT I COME WITH THE GUARANTEE OF FOOD SECURITY & A MASSIVE REDUCTION IN IMPORTS.
PUHLEASE! MOST AREAS ALLOCATED FOR PALM OIL HERE AREN'T EVEN ENVIRONMENTALLY SUITED FOR MASS YEILDS!
WELL, I CAN STILL FOOL THE AUTHORITIES INTO BELIEVING THAT I'M BETTER THAN SHIFTING CULTIVATION.
DELUDE YOURSELF! SHIFTING CULTIVATION STILL SEES OUR PEOPLE THROUGH, WHILE ALSO HOSTING FIVE TIMES MORE BIODIVERSITY THAN ANY MONO-CULTURE!
SO WHAT? I'M THE 'IN THING'!
YOU CALL GENOCIDE THE 'IN THING'?! YOU'VE DESTROYED WILDLIFE WHEREVER YOU'VE GONE, EXPLOITED INDIGENOUS COMMUNITIES & STRIPPED THEM OFF THEIR LAND RIGHTS!
YOU HAVEN'T CONQUERED COUNTRIES, MISTER. YOU'VE RAVAGED THEM. YOU'RE THE GREATEST MODERN ECOLOGICAL DISASTER! YOU ARE DOOM!
BUT THERE'S BIG MONEY RIDING ON ME, BABY!
BAHAHAHA! LOOK WHO'S 'DOOMED'!
TOUCHÉ.
Ronan

#11

THE WILD INDIAN SPRING
BLOOMING BOMBAX
MURMURATING STARLINGS
BLOOMING PALASH
FROLICKING ORIOLES
BLOOMING MAHUA
DRUNK SLOTH BEARS
YOU WERE TAGGED IN ONE PICTURE LAST NIGHT.

#12

POLLEN TRANSFER? IS THAT ALL OUR RELATIONSHIP WAS ABOUT?
UMM... YES.
AND ALL THOSE PROMISES? DID THEY MEAN NOTHING?
LOOK, YOU'RE OVERREACTING. THOSE WEREN'T PROMISES, JUST ALLOMONES.
LET ME EXPLAIN. I'M NOT A BEE. I'M A BEE ORCHID. THIS PART OF ME, THE LABELLUM, APPEARS LIKE FEMALE BEES AND EMITS ALLOMONES MIMICKING THEIR SCENT, TO ATTRACT MALE BEES TO ME.
AND THOSE MOMENTS OF FERVENT PASSION BETWEEN US?
IT WAS JUST PSEUDOCOPULATION, TO ENSURE I COULD TRANSFER MY POLLEN THROUGH YOU.
I TRUSTED YOU, RITA.
MY NAME IS NOT RITA.

#13
HAVE YOU BEEN ACCUSED OF BEING AN AWFUL HUGGER?
WHAT WILL IT TAKE FOR YOU TO GIVE A PROPER HUG?
DON'T WORRY, YOU'RE NOT ALONE!
MANY KINDS OF TREES ARE BAD HUGGERS TOO!
WE BELIEVE IN GIVING EACH OTHER OUR PERSONAL SPACE. THIS HAS MANY BENEFITS!
BY MAINTAINING A SPACE BETWEEN CANOPIES WE ENSURE THAT SUNLIGHT IS DISTRIBUTED EQUALLY, AND LEAF-EATING LARVAE DO NOT SPREAD EASILY.
IF YOU'RE STILL NOT CONVINCED THAT BEING A BAD HUGGER IS OKAY, THERE'S A FANCY TERM FOR IT—
'CANOPY SHYNESS'!
FORGIVE ME, I HAVE CANOPY SHYNESS.
Rohan

#14

COME WINTER, AND THE GHOST TREE STRIPS ITSELF OF ALL FOLIAGE, PROUDLY DISPLAYING ITS NAKED SPLENDOUR!
THE TREE REMAINS LEAFLESS FOR HALF THE YEAR, PRODUCING RED, VELVETY, STAR-SHAPED FRUITS THAT ATTRACT SEED DISPERSERS.
IT WILL NOT BE UNTIL MONSOON THAT LEAVES RE-APPEAR AND DRAPE THIS RADIANTLY BARE TREE THAT OUTSHINES EVERY OTHER TREE IN THE MOONLIGHT.
BOUDOIR MODELLING, HERE I COME!

#15

IT'S SUMMER AND THE RED SILK COTTON TREES HAVE BLOOMED ALL AROUND THE CITY...
HUNDREDS OF ROSY STARLINGS HAVE ARRIVED TO FEAST ON THE VERMILION-COLOURED FLOWERS...
THEIR FLOCKS PERFORM THE MOST SPECTACULAR DISPLAYS OF AERIAL ACROBATICS BEFORE THEY PERCH...
THE MOST STUNNING SCENES OF SUMMER ARE PLAYING LIVE IN THE URBAN BACKYARD...
AND THE MEDIA IS STILL OBSESSED WITH POLITICIANS & CELEBRITIES.
SWEEPS POLLS YET AGAIN!
WAR OF WORDS IN B TOWN!
SPOTTED IN RED HOT BIKINI!
TCH TCH TCH.
IT'S OKAY, I GUESS. ONE NEEDS SUPERIOR INTELLECT TO APPRECIATE ART.

#16

TODAY I BRING YOUR ATTENTION TO ONE OF THE BIGGEST THREATS TO INDIA'S FORESTS...
NO, NO, I'M NOT TALKING ABOUT THE ENVIRONMENT MINISTRY THIS TIME! I'M REFERRING TO AN INVASIVE AMERICAN SHRUB—
THE DREADED LANTANA. THIS SHRUB FOUND ITS WAY INTO INDIA IN THE EARLY 19th CENTURY AND NOW OCCUPIES 2/5th OF INDIA'S TIGER RANGE. THAT'S A SIGNIFICANT FRACTION OF THE 'GREEN COVER' WE'RE SO PROUD OF!
THIS PEST PLANT NOT ONLY COMPETES WITH NATIVE FLORA FOR RESOURCES, BUT IT ALSO COSTS INDIA A BOMB. PRECIOUS CONSERVATION FUNDS GET DIRECTED TOWARDS ITS ERADICATION ANNUALLY!
ECOLOGISTS PROPOSE COMPLETE UPROOTING AND CONTROLLED FIRES AS MEASURES TO GET RID OF LANTANA.
I PROPOSE THAT WE CLEAR LANTANA WITH THE SAME BLOODLUST WITH WHICH OUR GOVERNMENT CLEARS OUR FORESTS!
ROHAN

EIGHT

Wildlife Science and Conservation

#1

BIRD CONSERVAN
BIRD CONSERVAN
ALUMINIUM RINGS? REALLY? SHOW ME THE PLATINUM RANGE, SWEETHEART.

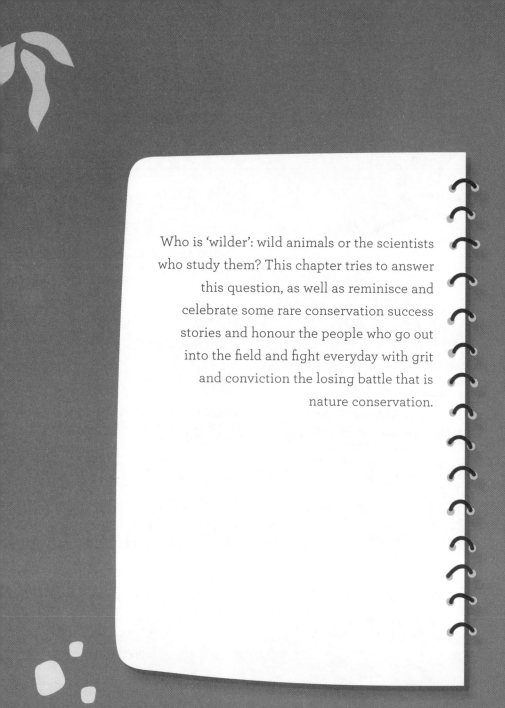

Who is 'wilder': wild animals or the scientists who study them? This chapter tries to answer this question, as well as reminisce and celebrate some rare conservation success stories and honour the people who go out into the field and fight everyday with grit and conviction the losing battle that is nature conservation.

#2

EVERY WINTER, WE, AMUR FALCONS EMBARK ON AN EPIC MIGRATION FROM EAST ASIA TO SOUTHERN AFRICA...
AND STOP OVER IN THE VILLAGES OF NORTH-EAST INDIA, IN NAGALAND, MANIPUR & ASSAM.
SOME YEARS BACK, TENS OF THOUSANDS OF US WERE KILLED AND SOLD AS MEAT IN LOCAL MARKETS.
BUT NOW, VILLAGERS HAVE FORMED COALITIONS TO PROTECT US FROM HUNTERS...
FRIENDS OF AMUR FALCONS
CELEBRATIONS ARE HELD TO MARK OUR ARRIVAL IN THESE VILLAGES...
STRUCTURES & PLACES HAVE BEEN NAMED AFTER US TO SPREAD AWARENESS...
AMUR FALCON LODGE
AMUR FALCON BEAUTY SALON
AMUR FALCON HOMESTAY
AMUR FALCON RESTAURANT
AND NO EFFORT IS SPARED TO ENSURE THAT WE HAVE A SAFE STAY IN INDIA.
I THINK THE NORTH-EAST JUST TRUMPED THE REST OF INDIA IN THE HOSPITALITY SECTOR!

#3

ONE SIMILARITY BETWEEN FIELD BIOLOGISTS AND DUNG BEETLES:

TIGER SCAT! YAYYY!!
TIGER SCAT! YAYYY!!

#4

DISADVANTAGES OF BEING RADIO-COLLARED

BEEP! BEEP!
MENU

#5
SOME 'WILD' WOMEN OF INDIA
J. VIJAYA (1959-87)
INDIA'S FIRST FEMALE HERPETOLOGIST. EXPLORED RAINFORESTS & CAVES ALL ALONE, RESEARCHING FOREST TURTLES BACK IN THE '80s.
PRERNA SINGH BINDRA
JOURNALIST & AUTHOR WHO STARTED WRITING ABOUT CONSERVATION WHEN THE MAINSTREAM MEDIA BARELY SPOKE ABOUT WILDLIFE.
APARAJITA DATTA
A PIONEER OF HORNBILL CONSERVATION IN INDIA. STARTED THE 'HORNBILL NEST ADOPTION PROGRAM' IN ARUNACHAL, AND CONVERTED MANY HUNTERS TO PROTECTORS.
DIVYA MUDAPPA
AN EXPERT ON HORNBILLS & SMALL CARNIVORES WHO HAS STUDIED THE RARE BROWN PALM CIVET IN THE WESTERN GHATS.
VIDYA ATHREYA
A LEOPARD SPECIALIST WHO HAS WORKED EXTENSIVELY TO STUDY & RESOLVE THE CONFLICT BETWEEN LEOPARDS & PEOPLE.
KIRAN PATHIJA
A FOREST GUARD IN GIR, WHO PATROLLED HER RANGE EVEN DURING HER PREGNANCY.
NANDINI VELHO
A BIOLOGIST WHO HAS BEEN WORKING EXHAUSTIVELY TO INVOLVE COMMUNITIES IN CONSERVATION IN THE NORTH-EAST, EFFECTIVELY BLENDING SCIENCE WITH COMMUNICATION & DESIGN.
TIASA ADHYA
A WILDLIFE BIOLOGIST & WRITER COMMITTED TO SAVING THE ENDANGERED FISHING CAT IN BENGAL FROM RETALIATORY KILLINGS & POACHING.
DON'T I ALWAYS KEEP TELLING YOU KIDS? MATRIARCHY MIGHT JUST SAVE THE WORLD SOME DAY.

#6

NEW RESEARCH BY INDIAN MARINE BIOLOGISTS CONDUCTED IN PALK BAY, OFF THE TAMIL NADU COAST, SHEDS NEW LIGHT ON CORAL BLEACHING...
THE STUDY SHOWS THAT CORALS FROM THE GENUS 'ACROPORA' ARE THE MOST SUSCEPTIBLE TO BLEACHING, WHILE THOSE FROM THE GENUS 'FAVITES', THE LEAST...
AND THAT BLEACHED CORALS SHIFT THEIR ASSOCIATION WITH SYMBIOTIC ALGAE, PREFERRING MORE HEAT-RESISTANT ALGAE DURING RECOVERY.
THESE FINDINGS WILL ENHANCE REEF TRANSLOCATION AND RECOVERY EFFORTS ACROSS INDIAN MARINE WATERS.
INDIAN MARINE BIOLOGISTS ARE MY NEW FAVITES... I MEAN FAVOURITES!

#7

send nudes
Nudibranch: Nembrotha sp.
Nudibranch: Hypselodoris sp.
Nudibranch: Chromodoris sp.

SEXTING WITH A MARINE BIOLOGIST.

#8

THE MADHYA PRADESH FOREST DEPARTMENT AND THE WILDLIFE CONSERVATION TRUST HAVE MADE HISTORY BY RADIO-TAGGING AND REHABILITATING TWO RESCUED INDIAN PANGOLINS!
FOREST DEPT.
THIS MEASURE IS BOUND TO CONTRIBUTE IMMENSELY TO STUDYING AND CONSERVING THE WORLD'S MOST TRAFFICKED MAMMAL!
HELLO, CONTROL ROOM, OVER. GUIDE ME TO THE NEAREST ANTHILL, OVER.
SORRY, THAT'S NOT HOW THE RADIO TAG WORKS, OVER.

#9

WE WANT A DIVORCE.
HMM.
DIVORCE ATTORNEY
ORNITHOLOGIST PAM RASMUSSEN WILL SPLIT YOU TWO INTO SEPARATE SPECIES.
DIVORCE ATTORNEY

#10

NINE TYPES OF ENVIRONMENTALISTS

THE ONLINE ACTIVIST
TAP TAP
RIDICULOUS how the use of fossil fuels shoots up EVERY @##%ing DAY!! are we so @##%ing stupid??!! when will mankind ever @##%ing learn?!! @##%@&*%#...
ONLINE ACTIVIST: TYPE 2
AAAND MY JOB FOR THE DAY IS DONE!
You shared SUPER ECO-FRIENDLY GIF's post on your friend's timeline.
THE FACT FILE
YOU MIGHT WANT TO GET YOUR FACTS RIGHT. THE RONGBUK GLACIER HAS RETREATED BY 33.04% AS OPPOSED TO THE 33% YOU JUST QUOTED.
THE ATHLETE
I'M RUNNING A MARATHON TO SAVE TIGERS, ALTHOUGH I'M NOT QUITE SURE HOW THEY'RE RELATED.
THE CAUSE CHAMPION
EARTH BUDDY
NO TO PLASTICS
YES TO SUSTAINABLE CONSUMPTION
CAUSE CHAMPION: TYPE 2
YOU MUST SWITCH TO SUSTAINABLE FABRIC.
HOW DO YOU KNOW THAT'S SUSTAIN-ABLE?
THE AD SAID SO.
THE OVERLY GREEN
HONEY, I GOT MYSELF STERILIZED! WE'LL NEVER BRING ANOTHER PARASITE INTO THIS PLANET!
THE VICIOUS VEGAN
YOU'RE NOT A VEGAN?! PLEASE GO DIE!
ACTUAL ACTIVISTS
HEY, DID ANYONE GET IN TOUCH REGARDING OUR CAMPAIGN?
YEAH, A POLITICIAN THREATENED TO ASSASSINATE US FOR MEDDLING WITH THE SAND MAFIA.

#11
PART-TIME JOBS OF A FIELD BIOLOGIST
EQUIPMENT TRANSPORT CARGO
PORNOGRAPHER
KHICHDI CHEF
SOCIAL MEDIA STRATEGIST
SAVE THE DHOLE
INVITE FRIENDS
OFFROAD STUNTS CO-ORDINATOR
LEECH EXTERMINATOR
FLICK
CAMPFIRE REGULATOR
SPY
CLICK
ELECTRICIAN
PATHOLOGIST
FIELD GEAR BRAND AMBASSADOR
DECATHLON
THE NORTH FACE
G-SHOCK
Quechua
VANGUARD
WILDCRAFT
Columbia
SALOMON

#12

EMBARRASSINGLY SHORT AMPLEXUS. MAY OR MAY NOT BE FOLLOWED BY INTERNAL FERTILIZATION.

HERPETOLOGY REVERSED

#13

Rhacophorus malabaricus AND *Raorchestes luteolus* ON *Oberonia recurva*.
TAXONOMY SOUNDS ALL LATIN TO ME.

#14

poink
Sniff
Sniff
I THINK HE'S DETECTED THE CAMERA TRAP.
ROMAN

#15

MY MOM THE BIRDWATCHER
COPPERSMITH BARBET!
NO, THAT'S A WOOD-PECKER.
ASHY PRINIA!
THAT'S A BABBLER.
BOOTED EAGLE!
STORK, MOM! USE YOUR BINOCULARS!
PARADISE FLYCATCHER!
NOPE. FORKTAIL.
UMM... SPOTTED DOVE?
THAT'S RIGHT!
WOOHOO!!! SIGN ME UP FOR THE BIG YEAR!

#16

TOURIST IN A CAVE:
UGH! BAT SHIT!
CHIROPTEROLOGIST IN A CAVE:
MMM... OH GUANO, YOU HEAVENLY ELIXIR!

#17

HOW MAN-MADE CANOPY BRIDGES HELP THE ENDEMIC & ENDANGERED LION-TAILED MACAQUES IN THE ANAMALAI HILLS
① CONNECT FRAGMENTED RAINFORESTS
② PREVENT ROADKILLS
③ CONNECT MACAQUE TERRITORIES DISRUPTED BY ROADS, PLANTATIONS AND URBANIZATION
④ THE PERFECT LOCATIONS FOR MACAQUE PRE-WEDDING PHOTOSHOOTS

#18

I... I... I CAN'T!
NINTENDO
huhuhu hu hu hu hu!
quack!
HIT
SCORE

A WILDLIFE CONSERVATIONIST PLAYS 'DUCK HUNT.'

SIMILARITIES BETWEEN DOLPHINS AND MARINE BIOLOGISTS
CREATURES OF EXTREME INTELLIGENCE
TALENTED WHISTLERS
ALWAYS SMILING
RIDICULOUSLY PHOTOGENIC
SPECIALIZE IN FLIPS
EXPERTS AT BURNING BODY BLUBBER
HAVE OCCULT UNDERWATER COMMUNICATION TECHNIQUES
EXTREMELY ELUSIVE
GMAIL
AUTO: ON-FIELD
Thanks for your mail. On-field until the end of the year. Sorry!
-ROHAN

#20

SOME NEWLY DISCOVERED SPECIES FROM 2019 CONGRATULATE INDIA'S BIOLOGISTS
THE IMPRESSED TORTOISE*
MIGHTY IMPRESSIVE, FOLKS!
CLAP CLAP
(* first record for India)
THE CRYING KEELBACK
FINALLY, A REASON TO SMILE!
20 NEW SPECIES OF 'IMPATIENS' BALSAM*
ALREADY IMPATIENT ABOUT NEXT YEAR'S DISCOVERIES!
(* between 2010 & 2019)
THE STARRY DWARF FROG
BIOLOGISTS ARE THE REAL STARS OF THE ANTHROPOCENE.
THE LAUDANKIA VINE SNAKE
THE WINE'S ON ME, GUYS!
THE AMBOLI BROOKIISH GECKO
MAY THE VAN DER WAAL'S FORCE BE WITH YOU!

#21

SOME NEW BIOLOGICAL BREAKTHROUGHS OF 2020 CONGRATULATE OUR BIOLOGISTS:
AENIGMACHANNIDAE, A NEW FAMILY OF BONY FISH FROM THE WESTERN GHATS
MAY YOU CONTINUE THIS 'AENIGMATIC' RUN NEXT YEAR!
MAGNIFICENT DWARF GECKO, A NEW LIZARD FROM KARNATAKA
MAGNIFICENT, AND NO DWARF FEAT!
Aloe trinervis, A NEW ALOE FROM RAJASTHAN
NEWS LIKE THIS MAKES MY SKIN GLOW!
BRANDED ROYAL BUTTERFLY, REDISCOVERED IN TAMIL NADU
I BRAND THIS A ROYAL CONTRIBUTION TO SCIENCE!
5 NEW AMPHIBIANS ADDED TO THE BIODIVERSITY OF MADHYA PRADESH, FROM PANNA TIGER RESERVE
LET'S CONTINUE LINKING BIOGEOGRAPHICAL DOTS, NOT RIVERS!
VAIBHAV'S PROTANILLA, A NEW SPECIES OF BLIND, UNDERGROUND ANT FROM NETRAVALI, GOA
THANKS FOR LEAVING NO EXCUSE FOR THE GOVERNMENT TO TURN A BLIND EYE TO GOA'S FORESTS!

#22
SOME SIMILARITIES BETWEEN LEOPARDS AND FOREST GUARDS
A SILENT BUT UBIQUITOUS PRESENCE ACROSS INDIA
BACKBONES OF THE INDIAN JUNGLE
24X7 VIGIL, EVEN IN THEIR SLEEP
EXPERTS AT CAMOUFLAGE & STEALTH
BETTER THAN GOOGLE MAPS AT NAVIGATING THE FOREST
SEASONED WEIGHTLIFTERS
MAKE THE MOST OF SCANTY RESOURCES
THEIR CONTRIBUTION TO THE HEALTH OF FORESTS IS OFTEN OVERLOOKED
EQUALLY AT HOME ON ANY TERRAIN
GLORIOUS WHISKERS!
THEY'RE BOTH IN DIRE NEED OF OUR SUPPORT.
ANTI-POACHING CAMP

#23
SOME SIMILARITIES BETWEEN OWLS AND WILDLIFE VETERINARIANS
TOUGH EXTERIOR, GENTLE HEARTS.
CRYPTIC FIELD PLUMAGE
USE THEIR INVENTORY WITH PIN-POINT ACCURACY
THE PERFECT MARRIAGE BETWEEN SKILL & INSTINCT
COMPLETE DAREDEVILS
SILENTLY AT WORK WHEN THE WORLD SLEEPS.
THEIR CONTRIBUTION TO CONSERVATION IS CRITICAL YET UNSUNG.
ROHAN

SOME WOMEN-LED CONSERVATION INITIATIVES IN INDIA

INSEASON FISH (DIVYA KARNAD):

PROMOTING SUSTAINABLE & ARTISANAL FISHERY & REDUCING BYCATCH, IN COLLABORATION WITH FISHING COMMUNITIES.

HARGILLA ARMY (PURNIMA DEVI BURMAN):

AN ALL-WOMAN TEAM IN ASSAM WORKING TO CONSERVE THE ENDANGERED HARGILLA (GREATER ADJUTANT STORK).

TERRA CONSCIOUS (PUJA MITRA):

WORKING IN CONJUNCTION WITH COASTAL COMMUNITIES IN GOA TO ESTABLISH ETHICAL MARINE ECO-TOURISM PRACTICES.

FRIENDS OF AMUR FALCONS (BANO HARALU):

ENSURING A SAFE PASSAGE FOR MIGRATING AMUR FALCONS IN NAGALAND BY SAVING THEM FROM HUNTERS.

THE LAST WILDERNESS (VIDYA VENKATESH):

COLLABORATING WITH FOREST DEPARTMENTS & TRIBAL COMMUNITIES TO MINIMIZE CONFLICT & SUPPORT ALTERNATIVE LIVELIHOODS.

GREEN HUB (RITA BANERJI):

A COMMUNITY & YOUTH-BASED FELLOWSHIP, TRAINING YOUNGSTERS FROM THE NORTH-EAST TO BE WILDLIFE FILMMAKERS.

MUD ON BOOTS (CARA TEJPAL, SANCTUARY ASIA):

EMPOWERING & SUPPORTING GRASSROOTS CONSERVATIONISTS ACROSS THE COUNTRY BY RAISING & ALLOCATING FUNDS & EQUIPMENT.

BLACK BAZA COFFEE (ARSHIYA BOSE):

ENABLING COFFEE GROWERS TO ADOPT FARMING PRACTICES THAT SUPPORT BIODIVERSITY.

WOW. PANTHRESSES LEADING THEIR OWN PACKS! THAT'S A FIRST.

#25

A CONSERVATION BIOLOGIST GOES FOR THERAPY

CAN YOU DESCRIBE WHAT YOU FEEL?
MISERY! DESPAIR! A CONSTANT THROBBING GRIEF!
THE FROG SPECIES I'VE BEEN TOILING TO CONSERVE IS DOWN TO FIVE INDIVIDUALS - ALL IN MY LAB.
THE GOVERMENT HAS NOW SANCTIONED A MINE AND A DAM AT MY FIELD SITE! MY CAREER IS AS DOOMED AS MY POST-DOCTORAL SPECIES!
MY PARTNER WORKING ON HORNBILLS LOST A WHOLE TRACT OF RAINFOREST TO PALM OIL MONOCULTURE! CAN YOU IMAGINE THE HAVOC THIS WOULD WREAK ON A THREATENED BIRD'S POPULATION?
ONE DAY THERE'S NEWS OF AN ENTIRE CORAL REEF DISAPPEARING; ANOTHER DAY WE HEAR OF A RACE OF RHINOS GOING EXTINCT!
WHAT KIND OF A WORLD ARE WE PASSING ON TO COMING GENERATIONS?! WHERE IS THAT ONE GLIMMER OF HOPE I COULD LATCH ON TO FOR SANITY'S SAKE?!
GUT-ROO-GOO.
AREN'T PIGEONS DOING REALLY WELL?
THEY'RE INVASIVE.
OH.

#26

BEING A MAN VS BEING A WOMAN IN CONSERVATION BIOLOGY AND RELATED VOCATIONS IN INDIA

SIR, I JUST PUBLISHED MY 10th PAPER!
EXCELLENT! LET ME WRITE YOU A POST-DOC RECOMMENDATION.
SIR, I JUST PUBLISHED MY 12th PAPER!
EXCELLENT! LUCKY NUMBER TO START THINKING OF HAVING KIDS NOW!
SIR, IF YOU COULD PLEASE SIGN MY FIELD PERMITS...
A PERMIT IS BUT A MERE PIECE OF PAPER! PLEASE HAVE SOME TEA, YOUNG MAN.
SIR, IF YOU COULD PLEASE SIGN MY FIELD PERMITS...
MADAM, HAVE SOME TEA, GO HOME, GET MARRIED AND BRING A HUSBAND ALONG. SOMEONE NEEDS TO BE RESPONSIBLE FOR YOUR SAFETY IN THE FOREST!
AND THEREFORE THIS STUDY CONCLUDES THAT IMMEDIATE INTERVENTION IS IMPERATIVE.
THY WISDOM HATH CHANGED MINE WORLDVIEW, SIRE! I HEREBY BESTOW KNIGHTHOOD UPON THEE!
AND THEREFORE THIS STUDY CONCLUDES THAT IMMEDIATE INTERVENTION IS IMPERATIVE.
WOAH, WOAH! WATCH THAT TONE, YOUNG LADY! DO YOU KNOW HOW MANY YEARS I'M SENIOR TO YOU?
HONOURED TO HAVE YOU AT OUR SYMPOSIUM, SIR.
THANKS.
HONOURED TO HAVE YOU AT OUR SYMPOSIUM, MA'AM. BY THE WAY, LOVELY PROFILE PICTURE! HOPE YOU DON'T MIND ME SAVING IT ON MY PHONE...HEHE. FREE FOR SOME DRINKS IN THE EVENING? I'M STAYING AT THE SAME HOTE...

NINE

Nature and **Governance**

SO WHAT DOES THE COP 25 MEAN FOR US POLAR ANIMALS?
#1
COPE 25.
WHAT?
COPE WITH FAFFING AND INACTION ON CLIMATE CHANGE FOR ANOTHER 25 YEARS OR SO.

If you find it difficult to cope with this chapter, please blame your government! Even as a zoonotic pandemic, which can be traced back to environmental destruction, shakes the planet out of its wits, governments all around the world find it difficult to comprehend that the time to use nature as an Environment Day photo-op is gone, and that economies must adopt conservation as a core principle to ensure that they are sustainable. The few exceptional governments that are on the path to sustainable development are celebrated in occasional cartoons.

#2

STOP! THIS IS A WETLAND, NOT A WASTELAND! W-E-T- WET, NOT WASTE!
GOVERNMENTS SUCK AT SPELLINGS.
CONSTRUCTIONS PVT. LTD.
CONSTRUCTIONS PVT. LTD.

#3

SOME ANIMALS FROM DESERTS AND ARID REGIONS THAT YOU WON'T FIND IN A FOREST:

THE SPINY-TAILED LIZARD
THE DESERT FOX
THE RED SAND BOA
THE DESERT SCORPION
THE SAW-SCAL VIPER
THE DESERT CAT
THE INDIAN DESERT JIRD
THE CHINKARA
THE CRITICALLY ENDANGERED GREAT INDIAN BUSTARD

ARID ECOSYSTEMS DON'T NEED 'REFORESTATION'! THE BARREN BRAINS OF OUR POLICY-MAKERS DO.

#4

TONIGHT ON NATURE TV NEWS, BELIZE BANS OIL DRILLING ALONG BARRIER REEFS!
TV
I'M GLAD IT'S NOT JUST US BRAIN CORALS WHO HAVE BRAINS HERE!
TV

#5

CHILE SAYS NO TO BILLION-DOLLAR MINE TO SAVE THE HABITAT OF HUMBOLDT PENGUINS!
GOVERNMENTS AROUND THE WORLD MUST LEARN FROM CHILE...
IN MY HUMBOLDT OPINION.

#6

POACHING?! WILDLIFE TRADE?! RIDICULOUS ALLEGATIONS!
WE DO EVERYTHING WE CAN TO CARE FOR OUR ANIMALS!
THAILAND TIGER TEMPLE
IT GETS VERY HOT AROUND HERE, YOU SEE, SO WE PUT THEM IN THE COLD STORAGE.

#7

IT'S FREEZING IN HERE, AND WE'RE STILL TALKING ABOUT GLOBAL WARMING!
NBC NEWS CALLED IT THE 'GREAT FREEZE', AND WE'RE STILL TALKING ABOUT GLOBAL WARMING!
HECK, THERE'S ICE IN MY FREEZER, AND WE'RE STILL TALKING ABOUT GLOBAL WARMING!
SO LET'S PULL OUT OF THE PARIS CLIMATE AGREEMENT, IMPOSE RESTRICTIONS ON THE EPA & CLIMATE SCIENTISTS, AND REMOVE EVERY MENTION OF CLIMATE CHANGE FROM THE WHITE HOUSE RECORDS...
BECAUSE CLIMATE CHANGE IS A BIG, FAT HOA...
SPLAT!
THUD!
SOMEWHERE IN THE ARCTIC...
MAN! THAT WAS SOME THROW!
IT JUST HAD TO BE DONE.

#8

CONSERVATION THE FIRST WORLD WAY
WELCOME TO SAVE THE EARTH DOT ORG! DO YOU KNOW THAT OVER-CONSUMPTION IS ACCELERATING CARBON EMISSIONS AND GLOBAL WARMING?
PLEASE SUPPORT OUR CAMPAIGN TO COMBAT CLIMATE CHANGE...
BY BUYING THIS COOL T-SHIRT!

#9

OUR PLANET CAN NO LONGER SUSTAIN GROWING HUMAN POPULATIONS!
DIDN'T YOU JUST HAVE YOUR THIRD CHILD?
LET ME EXPLAIN. THERE ARE CHILDREN... AND THEN THERE ARE WHITE CHILDREN...

#10

APEX PREDATORS ARE KEYSTONE SPECIES AND THE BACKBONES OF EVERY ECOSYSTEM! TOGETHER WE MUST ENSURE THEIR PROTECTION ALL AROUND THE WORLD!
HURRAY!!
IS THAT A WOLF?
I'LL BRB.
KACHUNK

#11

JOKOWI JOKES!
STAND-UP COMEDY BY THE INDONESIAN PRESIDENT!
SO... THE CLIMATE IS CHANGING! SEA LEVELS ARE RISING! JAKARTA IS DROWNING!
WHAT DO WE DO??!
SIMPLE! WE MOVE OUR CAPITAL TO INDONESIA'S COAL & OIL HUB— THE VERY CAUSES OF CLIMATE CHANGE!
AND, AND, AND— WE CONVERT FORESTS INTO VAST URBAN SPACES CALLED 'GREEN CITIES'!
HA HA HA
HA HA HA
LOL
SNORT
BY THE WAY, I'M SERIOUS.
ROHAN

#12

I'M THE TAPANULI ORANGUTAN, THE RAREST GREAT APE IN THE WORLD.
THE BATANG TORU HYDROPOWER PROJECT IN SUMATRA, INDONESIA, THREATENS TO DESTROY MY HABITAT AND WIPE MY SPECIES OUT!
AND NOT JUST ORANGUTANS, BUT OTHER ENDANGERED WILDLIFE SUCH AS SUMATRAN TIGERS & SUNDA PANGOLINS TOO WHICH SHARE OUR HABITAT.
IT DOESN'T END HERE! THE PROPOSED DAM LIES CLOSE TO A FAULT LINE, WHICH MAY LEAD TO AN EARTHQUAKE! THIS HAS CONVENIENTLY BEEN IGNORED IN ITS E.I.A!
DESPITE LOCAL & INTERNATIONAL CONDEMNATION, THE PROJECT ENJOYS FUNDING & BACKING FROM CHINA, AS PART OF ITS 'BELT AND ROAD' PROJECT...
OR IN OTHER WORDS, CHINA'S 'BELOW THE BELT' PROJECT.
ROHAN

#13

LOOKING AHEAD IS THE KEY TO ECONOMIC GROWTH. DON'T LOOK BACK NOW.
NATIONAL PARK
RAMAN

#14

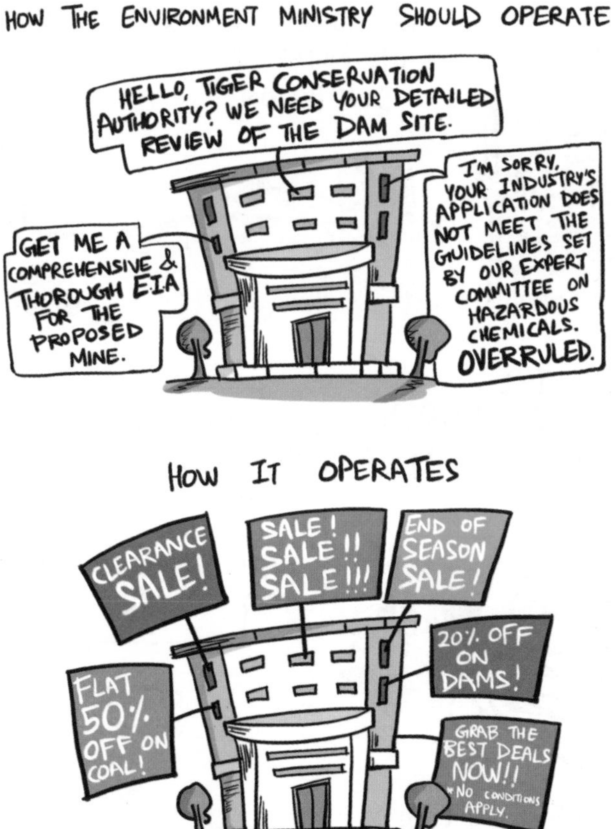
HOW THE ENVIRONMENT MINISTRY SHOULD OPERATE
HELLO, TIGER CONSERVATION AUTHORITY? WE NEED YOUR DETAILED REVIEW OF THE DAM SITE.
GET ME A COMPREHENSIVE & THOROUGH EIA FOR THE PROPOSED MINE.
I'M SORRY, YOUR INDUSTRY'S APPLICATION DOES NOT MEET THE GUIDELINES SET BY OUR EXPERT COMMITTEE ON HAZARDOUS CHEMICALS. OVERRULED.
HOW IT OPERATES
CLEARANCE SALE!
SALE! SALE!! SALE!!!
END OF SEASON SALE!
20% OFF ON DAMS!
FLAT 50% OFF ON COAL!
GRAB THE BEST DEALS NOW!! *NO CONDITIONS APPLY.
RAMAN

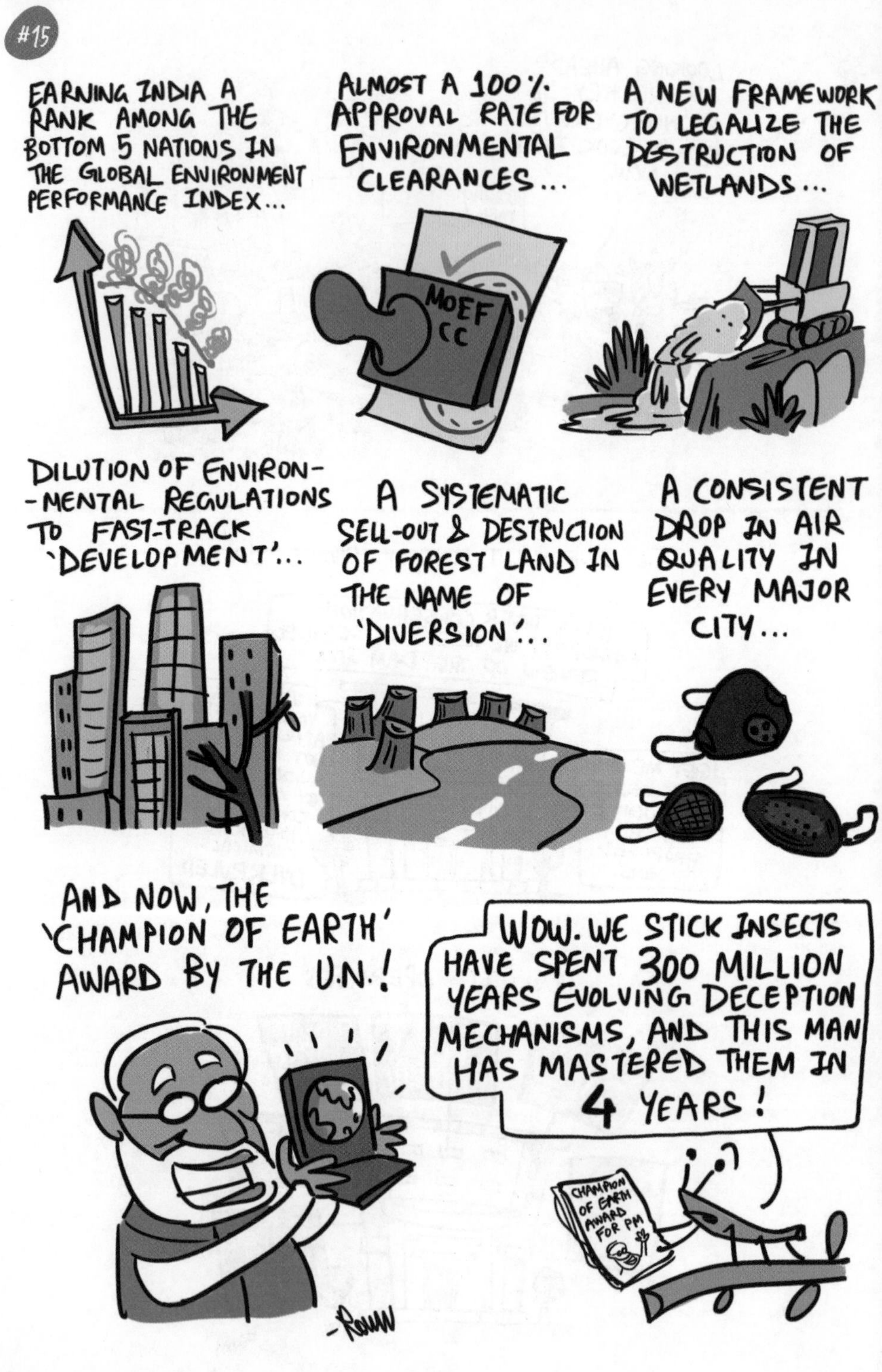
#15
EARNING INDIA A RANK AMONG THE BOTTOM 5 NATIONS IN THE GLOBAL ENVIRONMENT PERFORMANCE INDEX...
ALMOST A 100% APPROVAL RATE FOR ENVIRONMENTAL CLEARANCES...
MOEF CC
A NEW FRAMEWORK TO LEGALIZE THE DESTRUCTION OF WETLANDS...
DILUTION OF ENVIRON--MENTAL REGULATIONS TO FAST-TRACK 'DEVELOPMENT'...
A SYSTEMATIC SELL-OUT & DESTRUCTION OF FOREST LAND IN THE NAME OF 'DIVERSION'...
A CONSISTENT DROP IN AIR QUALITY IN EVERY MAJOR CITY...
AND NOW, THE 'CHAMPION OF EARTH' AWARD BY THE U.N.!
WOW. WE STICK INSECTS HAVE SPENT 300 MILLION YEARS EVOLVING DECEPTION MECHANISMS, AND THIS MAN HAS MASTERED THEM IN 4 YEARS!
CHAMPION OF EARTH AWARD FOR PM
-ROHAN

#16
THE MARINE LIFE OF MUMBAI PROTESTS AGAINST COASTAL RECLAMATION
THE OCTOPUS HAS TAKEN UP ARMS
THE STINGRAY AND THE JELLYFISH ARE CONDUCTING STINGS
ZOANTHIDS AND SPONGES HAVE FORMED A PROTEST CHAIN
MY CITY MY COAST
INTRUDERS GO BACK!
THE FILEFISH IS FILING A P.I.L
BIVALVES ARE SHOOTING AT RECLAMATION SITES WITH THEIR SIPHON PUMPS
Pew!
Pew!
AND THE SEA STAR IS USING HIS STAR POWER TO SPREAD THE WORD.
AGAINST COASTAL RECLAMATION
THE COAST IS NOT TO BE RECLAIMED FROM THE SEA. IT IS TO BE RECLAIMED FROM THE GOVERNMENT.
AGAINST COASTAL RECLAMATION

#17

THE AMAZON FIRES: A FLOWCHART

SOLD!
BBQ
#prayfor Amazon

#18

SCOTT MORRISON'S QUICK FIXES FOR CLIMATE CHANGE AND THE AUSTRALIAN BUSHFIRES

1) DENY ANY LINK BETWEEN THE FIRES & CLIMATE CHANGE.
NOT A CREDIBLE SUGGESTION.
2) TAKE A QUICK BREAK IN HAWAII.
I PROMISED THE KIDS!
3) EXPORT MORE COAL.
SEND IT ALL OUT!
4) WHY INCREASE FIREFIGHTING BUDGETS WHEN YOU CAN SIMPLY HUG FAMILIES OF DECEASED FIREFIGHTERS?
5) DISS GRETA THUNBERG.
NOT HERE TO IMPRESS PEOPLE OVERSEAS.
6) DON'T LET DYNAMIC & CAPABLE LEADERS FROM NEIGHBOURING COUNTRIES MAKE YOU INSECURE. LOOK THE OTHER WAY!
HMPF.
ZERO CARBON BILL

19

A HURDLE STANDS IN THE PATH OF 'DEVELOPMENT'!
CLEARANCE MAN, HELP!
EIA
SWISH
E
I
SLASH
A
SLOOSH!
BIG CORPORATIONS, HAVE NO FEAR...
EIA 2020 IS HERE!
EIA 2020
INTRODUCING POST-FACTO CLEARANCE:
OOPS! THEY ALREADY STARTED THE PROJECT!
NHAI
EXEMPTION OF PROJECTS FROM PUBLIC HEARING BY LABELING THEM 'STRATEGIC'.
I DECIDE WHAT'S BENEFICIAL FOR CITIZENS. NOT THE CITIZENS.
LEGITIMIZING ENVIRONMENTAL VIOLATIONS
NAUGHTY BOY!
STAMP
MoEF
FAIT ACCOMPLI
NOW THAT THE DAMAGE IS DONE, PLEASE PROCEED BY ALL MEANS!
EXEMPTION OF ROPEWAYS & WATERWAYS FROM OBTAINING CLEARANCE
PROMISE ME IT'S GREEN TRANSPORT.
PROMISE.
CRUSHING THE NATIONAL GREEN TRIBUNAL
NGT
AND MANY MORE!
DRAFT EIA 2020
~~ENVIRONMENT IMPACT ASSESSMENT~~
ENABLE INDIA'S AUCTION

#20
STAGGERING 90,000 HECTARES OF FOREST LAND 'DIVERTED' FOR DEVELOPMENT.
ANOTHER YEAR OF RIVERS RUNNING DRY, GOVERNMENT CLUELESS
CAN'T EVEN LINK TWO INCIDENTS LOGICALLY, AND THEY WANT TO LINK RIVERS.

#21
WHAT ROUTE DOES THIS BULLET TRAIN TAKE?
MUMBAI TO AHMEDABAD...
VIA SANJAY GANDHI NATIONAL PARK AND THANE CREEK...
BYPASSING THE WILDLIFE PROTECTION ACT.
THE
WILDLIFE
INDIA
1972

#22

SIR, AS AN ECOLOGIST WORKING IN THIS FOREST THAT YOU HAVE CLEARED FOR 'DEVELOPMENT', I HAVE A QUESTION.
HOW DO YOU PLAN TO 'DIVERT' AN AGE-OLD ECOSYSTEM, ITS ENDEMIC FLORA & FAUNA, ITS SYMBIOSIS WITH TRIBES DEVELOPED OVER CENTURIES, AND ITS CARBON SEQUESTRATION SERVICE?
TWO WORDS - COMPENSATORY AFFORESTATION.
JUST LOOK AT THESE MANICURED MARVELS! WE'VE NOT ONLY SUCCESSFULLY DIVERTED THE FOREST BUT ALSO INCREASED INDIA'S FOREST COVER!
THOSE ARE MONOCULTURE CASH CROPS & ORNAMENTAL TREES. HOW DO THEY CONTRIBUTE TO AN INCREASE IN FOREST COVER?!
BY CHANGING THE DEFINITION OF FOREST COVER. THAT'S HOW!
SO WHAT EXACTLY ARE YOU COMPENSATING FOR WITH COMPENSATORY AFFORESTATION?
ANY LOSS IN OUR P.R. RATINGS.
RUAN

#23
SOME SUGGESTIONS FOR THE ENVIRONMENT MINISTER
START LOOKING BEYOND THE MIRROR AND THIS PORTRAIT IN YOUR OFFICE FOR INSPIRATION.
BUY A BIRD ID GUIDE. START LOOKING OUT OF THE WINDOW.
GO ON NATURE TRAIL WITH KIDS. NOTE THEIR OBSERVATION DOWN
VISIT A COASTAL VILLAGE. GO FISHING WITH AN ARTISANAL FISHERMAN.
SNORKEL ALONG A CORAL REEF.
STAY AT A MAN-ANIMAL CONFLICT-PRONE TRIBAL HAMLET.
ACCOMPANY A FOREST GUARD ON A FOOT PATROL.
TREK WITH A FIELD BIOLOGIST.
RECAPITULATE ALL THAT YOU'VE LEARNED, IN THE SHADE OF A TREE.
NOW GO MAKE THIS OFFICE LIVE UP TO ITS NAME.
MINISTRY OF ENVIRONMENT, FORESTS & CLIMATE CHANGE

#24

LISTEN UP, EVERYONE! THE KEN-BETWA RIVER-LINKING PROJECT HAS BEEN CLEARED!
THUMP!
EVEN THE NATIONAL BOARD FOR WILDLIFE HAS GIVEN IT A NOD!
THUMP!
DESPITE THE FACT THAT IT'LL DROWN HUNDRED SQUARE KILOMETRES OF PRIME TIGER HABITAT IN PANNA TIGER RESERVE!
THUMP!
BUT DON'T YOU WORRY! JUST THUMP YOUR CHEST AND FEEL PROUD OF THE COUNTRY!
THUMP!
THIS EXERCISE MAKES EVEN THE MOST MONUMENTAL FAILURES BY THE GOVERNMENT SEEM LIKE RESOUNDING VICTORIES, AND MAKES YOU ETERNALLY POSITIVE ABOUT EVERY DIRE CONSEQUENCE! TRY IT!
THUMP!
THIS IS SO REASSURING.
THUMP!

TEN

Nature and **Us**

MEET SOME OF INDIA'S BIOLUMINESCENT CREATURES
MYCENA, A BIOLUMINESCENT FUNGUS.
BIOLUMINESCENT DINOFLAGELLATES, A TYPE OF PLANKTON
COMB JELLYFISH
DEEP SEA HATCHETFISH
SEVERAL SPECIES OF FIREFLIES AND GLOW WORMS
WE'VE BEEN CELEBRATING ECO-FRIENDLY DIWALI EVEN BEFORE DIWALI EXISTED!

#1

From how to be more sustainable to excelling in bed, nature has more lessons for us than Skillshare! And they're free of charge, provided we pause and look. The cartoons in this chapter speak about some ideas from nature that we can adopt in our daily lives. Some of the 'wrong' ones find their way in too, of course, which you may apply in your lives at your own risk.

#2

12 GREEN RESOLUTION SUGGESTIONS FROM WILD ANIMALS

#1: SWITCH TO PUBLIC TRANSPORT.

-The Cattle Egret

#2: START CARRYING YOUR OWN WATER BOTTLES TO AVOID USING PLASTICS.

-The Bactrian Camel

#3: ADOPT ORGANIC FASHION ACCESSORIES

-The Barasingh

#4: START CARRYING YOUR OWN SHOPPING BAGS.

(Corbiculae or pollen baskets)

-The Indian Hive Bee

#5: TURN TO VEGETARIAN ALTERNATIVES MORE OFTEN TO CUT DOWN ON MEAT.

-The Himalayan Black Bear

#6: USE YOUR BACKYARD TO GROW YOUR OWN FOOD.

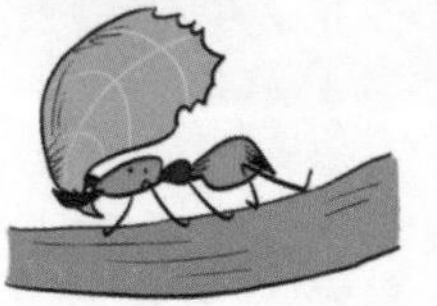

-The Leafcutter Ant

#7: BE INNOVATIVE WITH YOUR WASTE. CONVERT PLASTIC WASTE INTO HOME DECOR.

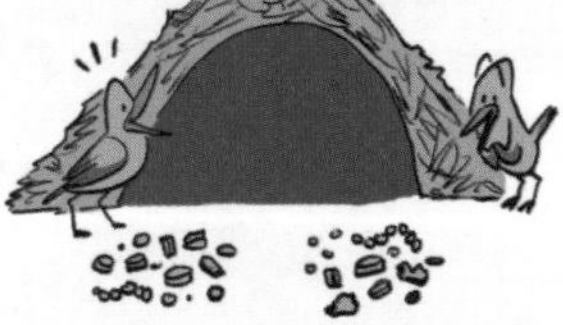

-The Vogelkop Bowerbird

#8: BE LESS OF A SUCKER. DITCH THE PLASTIC STRAW.

slurp!

-The Bonnet Macaque

#9: IF YOU'RE ONE OF THOSE WHO JUST CAN DO WITHOUT STRAWS, CARRY YOUR OWN REUSABLE STRAW.

-The Purple Sunbir

#10: BE A RESPONSIBLE FISH-EATER. AVOID THE FISH OF YOUR CHOICE DURING ITS BREEDING MONTHS.*

-The Smooth-coated Otter

*Refer to Knowyourfish.org.in

#11: DITCH CHEMICAL COSMETICS. SWITCH TO ORGANIC MAKE-UP & DYES.

-The Lesser Flamingo

#12: QUIT BEING A SNOB ABOUT SECOND-HA CLOTHING. IT'S SUSTAINABLE FASHION!

-The Hermit Cra

#3
11 WAYS TO SEDUCE YOUR MATE - A COURTSHIP GUIDE BY WILD ANIMALS
KNOCK THAT MALE FRIEND OF HERS DOWN. DESTROY HIM. MAKE SURE SHE'S WATCHING.
-The Musk Ox
GIFT HER A FISH. DO NOT LET HER EAT IT UNLESS SHE LETS YOU MOUNT.
-The Tern
FLASH HER.
-The Peacock
INITIATE A DUET. INVITE HER TO SING THE FEMALE LINES.
-The Sarus Crane
SET THE DANCEFLOOR ABLAZE.
-The Magnificent Riflebird
ADVERTISE YOUR PHEROMONES. PISS ALL OVER HER.
-The Porcupine
sniff
mmm!
SPRUCE UP YOUR LIVING ROOM BEFORE YOU INVITE HER HOME.
-The Bowerbird
INFLATE YOUR NECK INTO A HEART-SHAPED BALLOON.
-The Frigatebird
BE CHIVALROUS. GENTLY GROOM HER, EATING THE TICKS ON HER ARMPITS.
-The Macaque
Sigh!
TICKLE HER FACE WITH YOUR STREAMERS.
-The Twelve-wired Bird of Paradise
giggle!
END IT WITH A HICKIE ON HER NAPE.
-The Tiger
-Rouan

#4

AN INDIAN WILDLIFE VIDEO-CONFERENCE
BENGAL TIGER
HEY, EVERYONE! ALL SET?
GREATER FLAMINGO
UMM, MORE OR LESS.
BLACK-NAPED HARE
CAN ANYONE TELL ME HOW TO PUT THESE ON?
SLOTH BEAR
DON'T MIND THE HIC! HICCUPS, GUYS. IT'S MAHUA SEASON.
NARCONDAM HORNBILL
PLEASE EXCUSE THE ___ LAG. I LIVE ON A ___ REMOTE ISLAND___.
JUNGLE BABBLER
SORRY, MY FAMILY IS A LITTLE CHATTY.
CHEK!
CHEK!
CHEK!
CHEK!
BRAHMINY SKINK
SORRY, IT'S A LITTLE DARK UNDER-GROUND.
GOOSE BARNACLES
THE SIGNAL'S WEAK HERE.
MOVE TO ANOTHER SPOT.
WE'RE SESSILE.
MOTTLED WOOD OWL
WHOSE BRIGHT IDEA WAS IT TO HAVE A VIDEO CALL AT ELEVEN ****ING A.M.?!
SIGH. OKAY. LET'S JUST DO THIS TOMORROW.
ROHAN

#5
TIPS FOR AN ECO-FRIENDLY DIWALI FROM WILD ANIMALS
LET THERE BE LIGHT. JUST LIGHT.
-The Fireflies
DONATE OLD CLOTHES.
-The Hermit Crab
FEAST ON LOCALLY SOURCED SWEETS.
-The Honey Buzzard
USE BIODEGRADABLE MATERIAL FOR RANGOLI.
-The Signature Spider
USE EARTHENWARE INSTEAD OF SYNTHETICS.
-The Potter Wasp
SHARE GOODIES WITH YOUR NEIGHBOUR.
-The Grey Langur
KEEP NOISE TO A MINIMUM.
-The Barn Owl
MOULT INTO THE SAME DAZZLING ETHNIC SUIT FROM LAST YEAR!
-The Paradise Flycatcher

#6
TIPS FOR A GREEN CHRISTMAS BY INDIAN WILD ANIMALS
BE CREATIVE. 'REGENERATE' DECORATION FROM LAST YEAR INSTEAD OF BUYING AGAIN.
-The Indian Sea Star
MAKE HANDMADE GIFTS INSTEAD OF SHOPPING ONLINE. (DOESN'T NECESSARILY HAVE TO BE A BALL OF SALIVA)
-The Scorpionfly
IF YOU HAVE TO SHOP, CARRY YOUR OWN BAG.
-The Spot-billed Pelican
THE BEST GIFTS ARE THE ONES PRESENTED UNWRAPPED.
-The Blue-eared Kingfisher
COOKED A BIT TOO MUCH? CALL THE GANG OVER!
-The Long-billed Vulture
LOOKING FOR A CHRISTMAS CAKE? SUPPORT A LOCAL BAKER INSTEAD OF BUYING FROM A CONGLOMERATE.
-The Dung Beetle

#7

ANIMAL BABIES TELL US WHY THEIR MOMS ARE THE BEST MOMS

SHE WAS PREGNANT WITH ME FOR 22 MONTHS, THE LONGEST MAMMALIAN PREGNANCY EVER!

THE AFRICAN ELEPHANT

SHE NURSED ME FOR 7 YEARS, AND SHE MAKES ME A NEW BED EVERY SINGLE NIGHT!

THE ORANGUTAN

SHE LOST 7 POUNDS EVERY DAY WHILE NURSING ME!

THE HARP SEAL

SHE WAS SO STARVED LOOKING AFTER 200,000 OF US THAT SHE HAD TO EAT HER OWN ARM!

THE GIANT PACIFIC OCTOPUS

SHE LET ME EAT HER INSIDE OUT!

THE SEA LOUSE

#8

FABULOUS FATHERS OF INDIA

THE GREAT HORNBILL

SINGLE-HANDEDLY PROVIDES FOR BOTH HIS MATE AND HIS CHICKS, WHO STAY LOCKED TO ESCAPE PREDATORS.

THE LESSER FLAMINGO

FEMINIST TO THE CORE. SHARES ALL HOUSE-KEEPING DUTIES EQUALLY.

THE GHARIAL

GUARDS A CRÈCHE OF HUNDREDS OF HATCHLINGS, EVEN SERVING AS A BASKING STATION FOR THEM.

THE LITTLE GREBE

CARRIES HIS CHICKS ON HIS BACK TO KEEP THEM SAFE FROM PREDATORS.

THE PHEASANT-TAILED JACANA

A SINGLE DAD. HIDES HIS CHICKS UNDER HIS WINGS ON SPOTTING A THREAT.

THE SPOTTED SEAHORSE

DOESN'T CARE A JOT ABOUT GENDER STEREOTYPES. GETS PREGNANT HIMSELF!

#9

MOUNTAINEERING TIPS FROM HIMALAYAN ANIMALS
INVEST IN FOOLPROOF INSULATION.
-The Pallas' Cat
TAKE YOUR STAMINA-BUILDING REGIME SERIOUSLY.
-The Kiang
GET THE RIGHT FOOTWEAR FOR A STURDY GRIP.
-The Himalayan Tahr
STOCK UP ON FOOD SUPPLIES.
-The Himalayan Marmot
LISTEN TO THE MOUNTAINS MORE THAN YOU SPEAK.
-The Tibetan Lynx
YOU CANNOT FLY HIGH WITHOUT EXERCISING YOUR PECTORALS.
-The Black-necked Crane
NEVER DISCOUNT THE IMPORTANCE OF TEAMWORK IN CONQUERING A SUMMIT.
-The Brahminy Shelduck
A TRUE MOUNTAINEER IS HIS OWN PORTER.
-The Snow Leopard
AND A TRUE MOUNTAINEER NEVER LEAVES BEHIND ANY LITTER.
-The Bearded Vulture

#10

SOME WILD TIPS TO BEAT THE SUMMER HEAT
LICK YOUR ARMS WET. -The Red Kangaroo
PAMPER YOURSELF A BIT. WALLOW IN SLUSH. -The Water Buffalo
GROW A BUSHY TAIL & USE IT AS A HAT. -The Cap Ground Squirrel
DITCH YOUR SISSY SUNSCREEN FOR A UV-PROOF MUDBATH. -The Asian Elephant
BURY YOURSELF. -The Brahminy Blind Snake
RADIATE EXCESSIVE BODY HEAT THROUGH YOUR EARS. -The Fennec Fox
HOLD YOUR PISS THROUGHOUT SUMMER. -The Dorcas Gazelle
SOAK YOUR BREASTS IN WATER TO ABSORB MOISTURE. -The Painted Sandgrouse
PISS ALL OVER YOUR LEGS. -The White-backed Vulture

#11

WILDLIFE QUARANTINE SKILL-SHARING WEBINARS
SELF-ISOLATION & ZEN WITHIN A PINE CONE
Speaker: Indian Pangolin
SEALING YOURSELF SHUT WITH YOUR OWN SH%T
Speaker: Helmeted Hornbill
PORTABLE QUARANTINING: CARRYING YOUR HOME WITH YOU
Speaker: Hermit Crab
METAMORPHOSING MEDITATION
Speaker: Rose Butterfly Caterpillar
QUARANTINE CACHING: STOCKING UP FOR AN UNDERGROUND LIFE
Speaker: European Mole
HOME YOGA & WEIGHTLESS WORKOUTS
Speaker: Rock Agama
EMBROIDERING WITH YOUR CHILD'S SECRETIONS
Speaker: Weaver Ant
NAPPING THROUGH POINTLESS WEBINARS: AN EMERGING ART
Speaker: Three-toed Sloth

#12
SIMILARITIES BETWEEN THE LANGUR MOM AND MY MOTHER
LOVELY EYELASHES
GREAT CLIMBERS
DUST DUST
EXPERT MULTITASKERS
PLUCK
COMMUTING WITH KIDS WAS NEVER A PROBLEM
THE BEST GROOMERS
wiggle wiggle
NICE, LONG TAILS
LETS RANDOM AUNTS MANHANDLE ME
EFFECTIVE, TIMELY ALARM CALLS
KHAKK!
PSST! OUT WITH YOUR LADY FRIEND TO THE LIVING ROOM. YOUR DAD'S AT THE GATE.
-ROHAN

#13
GREEN TIPS FOR THE ECO-CONSCIOUS OFFICE-GOER
BY
CAPTAIN SUSTAINO
TAKE THE SHORTEST ROUTE TO WORK.
PROMOTE AN ECO-FRIENDLY DRESS CODE.
BURN NO EXTRA FUEL TO MEET YOUR DATE. DATE COLLEAGUES ONLY.
USE OLD DVDS INSTEAD OF PAPER FOR PAPER FIGHTS.
HIBERNATE COMPUTERS PERIODICALLY.
HARVEST RAINWATER DURING BREAKS.
PROMOTE GREEN CUISINE IN THE OFFICE CAFETERIA.
SAY NO TO PAPER CUPS. DRINK FROM THE COFFEE MACHINE.
USE REJECTED CVs INSTEAD OF TOILET PAPER.
GREEN UP YOUR OFFICE FURNITURE.
BOSS
PRINT AS LITTLE AS POSSIBLE.
CONFERENCE

ELEVEN

The New Normal

DID YOU KNOW THAT CHINA HAS BANNED ALL WILDLIFE TRADE AFTER THE CORONAVIRUS OUTBREAK?
WOW! IT'S TIME WE UPDATED OUR DEFENCE MECHANISM AGAINST POACHING.
I HAVE CORONAVIRUS! LET'S PLAY TAG!

#1

What started as an end-of-the-world alarm in Wuhan has now become the new normal globally. The COVID-19 pandemic has had its ups and downs, not just for us human beings but also for the natural world. The cartoons in this brief, final chapter explore the impacts of two dreadful viruses combined—COVID and humans—on the environment.

#2

BREAKING NEWS!
CORONAVIRUS SPREADS, INDUSTRIAL GROWTH AT AN ALL-TIME LOW.
BREAKING NEWS!
CRUISE & AVIATION INDUSTRIES TAKE MASSIVE BEATING AS PANDEMIC ESCALATES.
BREAKING NEWS!
CHINA BANS WILDLIFE TRADE IN THE WAKE OF PANDEMIC, SOUTH-EAST ASIAN COUNTRIES TO FOLLOW SUIT.
MY, MY! WHAT'S THE SECRET BEHIND THIS NEWFOUND GLOW?
SHH! DON'T JINX IT.

#3

AND YOU THOUGHT THE PANDEMIC IS FINALLY TEACHING MANKIND A LESSON.

#4

LISTEN TO THIS!!
OKAY, YOU FIRST.
SO, THE COMMUNIST CHINESE GOVERNMENT, FAILING TO TACKLE THE FAMINE IN THE '70s, LEGALIZED PRIVATE FARMING OF WILDLIFE, WHICH HAS WELL-ESTABLISHED LINKS WITH THE CORONAVIRUS PANDEMIC OF TODAY!
AND THE UNCONTROLLED DESTRUCTION OF NATURAL RESERVES DRIVEN BY CAPITALISM AROUND THE WORLD IS EXACERBATING THE SPREAD OF NEW PATHOGENS, BY INCREASING STRESS LEVELS IN CARRIER ANIMALS AND BRINGING THEM IN CONTACT WITH HUMANS.
WHICH MEANS THAT GOVERNMENTS AROUND THE WORLD HAVE A MUCH LARGER RESPONSIBILITY BEYOND CURBING THE PANDEMIC...
YES. ACKNOWLEDGING THAT EXISTING METHODS OF GOVERNANCE ARE FUNDAMENTALLY FLAWED, AND DESIGNING AND IMPLEMENTING NEW GOVERNANCE MODELS THAT HAVE ENVIRONMENT PROTECTION AT THEIR CRUX.
BUT WHAT DO YOU THINK COULD PROMPT IMMEDIATE, COLLECTIVE, GLOBAL ACTION OF THIS MAGNITUDE?
HMM.
DO YOU HAVE AN INTROSPECTIVE PANIC-INDUCING VIRUS IN YOUR INVENTORY THAT WE COULD TRANSMIT NEXT?

#5

LINKS BETWEEN CORONAVIRUS AND WILDLIFE MARKETS HAVE PUSHED THE CHINESE GOVERNMENT TO ENFORCE A BAN ON WILDLIFE TRADE.
BUT ONE OF THE VERY REASONS FOR THIS WILDLIFE TRADE— TRADITIONAL CHINESE MEDICINE— IS NOW BEING TOUTED AS A CURE FOR CORONAVIRUS!
TCM
BACKED BY PRESIDENT XI JINPING HIMSELF, TCM, WHICH OFTEN CONTAINS PARTS FROM ANIMALS KNOWN TO BE CARRIERS OF CORONAVIRUS, IS NOW FINDING NEW BUYERS IN CHINA AND LAOS!
TCM
TCM HAS BEEN RESPONSIBLE FOR ENDANGERING HUNDREDS OF SPECIES WORLDWIDE, SUCH AS RHINOS, HORNBILLS, GECKOS, TIGERS & PANGOLINS.
AT A TIME OF NATIONAL CRISIS, WHEN THE GOVERNMENT SHOULD BAN THE USE OF WILDLIFE IN A POINTLESS PSEUDOSCIENCE, IT IS DOING JUST THE OPPOSITE, PUTTING EVEN MORE WILDLIFE AND ITS OWN PEOPLE IN PERIL.
I HAVE JUST ONE WORD FOR THE CHINESE GOVERNMENT—
AA... AAA... AAAA
ACHHOOO!

#6

DEAR RUMOUR-MONGERS & WHATSAPP MILITIA; BATS DON'T GIVE YOU CORONAVIRUS.
YOUR ILL-TREATMENT OF BATS IS WHAT GIVES YOU VARIOUS DISEASES.
HOW? FOR EXAMPLE, DEFORESTATION & HABITAT LOSS COULD ELEVATE OUR STRESS LEVELS, WEAKEN OUR IMMUNE RESPONSE TO PATHOGENS, AND MAKE TRANSMISSION EASIER.
OR IF WE'RE HANDLED OR HUNTED & EATEN, YOU GET THE PATHOGENS WE CARRY.
SO THE SIMPLE WHATSAPP MESSAGE YOU SHOULD INSTEAD BE CIRCULATING IS THIS:
TREAT BATS WELL & LET US BE; AND YOU REMAIN INFECTION-FREE.
WHILE OF COURSE, ENJOYING OUR FREE SERVICES LIKE SEED DISPERSAL & PEST CONTROL.
PING!
"WORSHIP BATS TO PREVENT EPIDEMICS". -DOCTORS.
WHY MUST YOU ALWAYS OVERREACT, WHATSAPP?

#7

SHARKS ARE IN PERIL YET AGAIN. NOT FOR OUR FINS THIS TIME, BUT FOR THE SQUALENE IN OUR LIVER, TO BE USED IN COVID VACCINES.
FUNNILY FOR HUMANS, THE SOLUTION TO A PANDEMIC CAUSED BY ENVIRONMENTAL PLUNDER IS EVEN MORE ENVIRONMENTAL PLUNDER.
DESPITE THERE BEING OTHER HARMLESS SOURCES OF SQUALENE, LIKE VEGETABLE OILS & YEAST, THE PHARMA INDUSTRY IS EYEING US SHARKS AS A QUICK FIX.
THERE'S A QUICKER FIX IN MY MIND, NOT JUST FOR THE PANDEMIC, BUT FOR ALL ENVIRONMENTAL WOES OF PLANET EARTH. MAY I?
SQUALENE OCCURS IN HUMANS TOO.
JUST SAYING.

#8

LEARN PHYSICAL ISOLATION AND SOLITUDE FROM THE MASTERS

24 HOURS A DAY AREN'T ENOUGH FOR ME TO ENJOY MY OWN COMPANY.

-The Eurasian Lynx

I RESTRICT COMPANY TO PEOPL WHO LIKE MY STENCH, i.e. ME

-The Striped Skunk

TOO MUCH OF A PRICK FOR ANYONE TO HANG OUT WITH.

-The Indian Crested Porcupine

TOO INTIMIDATING A REPUTATIO FOR ANY MAN TO SWIPE RIGH

-The Black Widow Spider

RELATIONSHIP STATUS: UNDERGROUND UNTIL MONSOON.

-The Indian Purple Frog

NETFLIX & ~~CHILL~~ PARTHENOGENES

N

-The Brahminy Blind Snake

Acknowledgements

I owe a debt of gratitude and a session of social grooming to:

My parents, Sulabha and Ashit, for bearing with my ever-fluctuating career interests, for standing by my eventual decision to become a cartoonist and for supporting me immensely.

My brother, Rohit, a talented writer and bat biologist, for being my best birding companion and my worst critic.

My wife, Rithika, for her superlative taste in cartoons (and cartoonists) and for helping me refine my work on countless occasions with her wit and critique.

Cartoonists who have inspired and influenced my work: Bill Watterson, Patrick McDonnell, Gary Larson, Mark Parisi, R.K. Laxman and Nina Paley.

My brigade of mentors, friends and family who have had my back: Bittu Sahgal, Bikram Grewal, Prerna Bindra, Nandini Velho, Tina Fernandes, Roanit Fernandes, Pritha Dey, Dipsha Kriplani, Dhwani Chandel, Pallavi Talware, Munmun Dhalaria, Sejal Mehta, Bijal Vachharajani, Chemudupati Samyukta, Barkha Nagpal, Swati Sani, Tarique Sani and Anuradha Paul.

All the dogs in my life: Naughty, Rabie, Champak, Chhoti, Laxman, Kamla, Sunny, Nargis, Fifi, Srishti, Sakshi, Chandni and Wall-E, for helping me keep my sense of humour intact in an increasingly dreary world.

The publishers of my columns for believing in my work despite the ever-shrinking space for independent comics in newspapers and multimedia: Sree Nandy (*Saevus*); Soity Banerjee, Aditi Sengupta and Veena Venugopalan (BL Ink); Tinaz Nooshian, Gitanjali Chandrashekharan and Prutha Bhosale (*Sunday Mid-Day*); Vaishna Roy (*The Hindu*); Lalitha Suhasini and Vinutha Mallya (*Pune Mirror*); Ricky Kej, Neha Dara and Megha Moorthy (RoundGlass); Ayse Dincer, Gozde Polatkal and Derya Sahhuseyinoglu (Arastirmaci Cocuk);

Vinatha Vishwanathan (*Chakmak*); Radha Rangarajan (Nature In-Focus); Samantha Vuignier (Cartoon Collections) and Sheena Wolf (Gocomics). If they ever need someone to make pictures to offend their bosses or in-laws, they know who to call.

To Premanka Goswami, Devangana Dash, Parag Chitale and Aslesha Kadian at Penguin Random House India, for putting up with my whims, letting me know that an engaging comics compilation *cannot* be 800 pages long and for bringing this book to life.

My hometown, Nagpur, also known as 'The Tiger Capital of the World', for providing me with the most wonderful wildlife at my doorstep (often literally).

Lastly, and obviously, wildlife, for mesmerizing me and for captivating my imagination, right from the Arctic Terns on their way to either of the poles to the jumping spider feasting on the mosquitoes in my bathroom (the spider, of course, gets a hand-delivered signed copy).

AH!
WHAT'S HE SO EXHILARATED ABOUT?
COMPILING A BOOK OF CONSERVATION CARTOONS AND FEELING LIKE AN ECO-WARRIOR IN HIS HEAD, I GUESS...
ECO-WARRIOR? YEAH, RIGHT! ALL THOSE CARTOONS WERE DRAWN ON A TABLET & COMPUTER THAT RUN ON THERMAL POWER!
WHILE GUZZLING COFFEE AND HOGGING ON CHORIZOS & PEANUT BUTTER...
AND USING A SMARTPHONE MADE FROM COBALT & LITHIUM, FOR 'OUTREACH'!
HAHAHAHA!
WAIT. LET ME CHECK THE CARBON FOOTPRINT OF YOUR FAVOURITE TREATS...
PSST. SOMEONE'S EGO HERE IS MORE FRAGILE THAN THE LARSEN-C ICE SHELF!